DNA Cloning 1

P9-AGG-278

DATE DUE			
NO 12 98			
DE 7 '98			
AP 20 '99			
MY 27 99			
NO 5 99			
JE 1 0 '00			
FE 8 '01			
AP 2 '01			
AP 1 8 '01			
NO 1 3 '01			
DE 7 '01			
MY 1 6 '02			
NO 9 '02			
FE 1 3 '04			
DE 17 '05			
JE 1 0 08			

The Practical Approach Series

SERIES EDITORS

D. RICKWOOD
Department of Biology, University of Essex
Wivenhoe Park, Colchester, Essex CO4 3SQ, UK

B. D. HAMES
Department of Biochemistry and Molecular Biology
University of Leeds, Leeds LS2 9JT, UK

Affinity Chromatography
Anaerobic Microbiology
Animal Cell Culture
  (2nd Edition)
Animal Virus Pathogenesis
Antibodies I and II
Basic Cell Culture
Behavioural Neuroscience
Biochemical Toxicology
Biological Data Analysis
Biological Membranes
Biomechanics—Materials
Biomechanics—Structures and
  Systems
Biosensors
Carbohydrate Analysis
  (2nd Edition)
Cell–Cell Interactions
The Cell Cycle
Cell Growth and Division
Cellular Calcium
Cellular Interactions in
  Development
Cellular Neurobiology

Centrifugation (2nd Edition)
Clinical Immunology
Computers in Microbiology
Crystallization of Nucleic Acids
  and Proteins
Cytokines
The Cytoskeleton
Diagnostic Molecular Pathology
  I and II
Directed Mutagenesis
DNA Cloning 1: Core Techniques
DNA Cloning 2: Expression
  Systems
DNA Cloning 3: Complex
  Genomes
DNA Cloning 4: Mammalian
  Systems
Drosophila
Electron Microscopy in Biology
Electron Microscopy in
  Molecular Biology
Electrophysiology
Enzyme Assays
Essential Developmental Biology

DNA Cloning 1
Core Techniques
A Practical Approach

Edited by

D. M. GLOVER

Cell Cycle Genetics Research Group, CRC Laboratories,
Department of Anatomy and Physiology, Medical Sciences Institute,
The University, Dundee DD1 4HN, UK

and

B. D. HAMES

Department of Biochemistry and Molecular Biology,
University of Leeds, Leeds LS2 9JT, UK

IRL PRESS
——at——
OXFORD UNIVERSITY PRESS
Oxford New York Tokyo

Valton Street, Oxford OX2 6DP

New York
d Bangkok Bombay
n Dar es Salaam Delhi
ong Istanbul Karachi
Iras Madrid Melbourne
robi Paris Singapore
Taipei Tokyo Toronto

and associated companies in
Berlin Ibadan

Oxford is a trade mark of Oxford University Press

Published in the United States
by Oxford University Press Inc., New York

© Oxford University Press, 1995

First printed 1995
Reprinted 1996

A catalogue record for this book is available from the British Library

Library of Congress Cataloging-in-Publication Data
DNA cloning: a practical approach / edited by D.M. Glover and
B.D. Hames. —2nd ed.
p. cm.– (Practical approach series; v. (148, 149, 163, 164))
Includes bibliographical references and indexes.
Contents: v. 1. Core techniques —
1. Molecular cloning—Laboratory manuals. 2. Recombinant DNA—
Laboratory manuals. I. Glover, David M. II. Hames, B. D.
III. Series.
QH442.2.D59   1995   574.87'3282—dc20   94-26740

ISBN 0 19 963477 7 (Hbk)
ISBN 0 19 963476 9 (Pbk)

Printed in Great Britain by Information Press Ltd, Eynsham, Oxon.

Preface

It is a decade since the first editions of *DNA Cloning* were being prepared for the Practical Approach series. It is illuminating to look back at those volumes and reflect how the field has evolved over that period. We have tried to distil out of those earlier volumes such techniques that have withstood the test of time, and have asked many of the former contributors to update their chapters. We have also invited several new authors to write chapters in areas that have come to the forefront of this invaluable technology over recent years. The field is, however, far too large to cover comprehensively, and so we have had to be selective in the areas we have chosen. This has also led to each of the individual books being focused on particular topics. This having been said, Volume 1 covers core techniques that are central to the cloning and analysis of DNA in most laboratories. Volume 2 on the other hand turns to the systems used for expressing cloned genes. Inevitably the descriptions of these techniques can be supplemented by reference to other cloning manuals such as *Molecular cloning* by Sambrook, Fritsch, and Maniatis (Cold Spring Harbor Laboratory, New York, 1989) as well as other books in the Practical Approach series. Volume 3 examines the analysis of complex genomes, an area in which there have been many important developments in recent years, both in the description of new vectors, and in the strategic approaches to genome mapping. Again, companion volumes such as *Genome analysis* can be found in the Practical Approach series. Finally, in Volume 4 we look at DNA cloning and expression in mammalian systems; from cultured cells to the whole animal.

This volume begins with a description of the techniques for the transformation of *E. coli* from Doug Hanahan's laboratory and is a revision of the chapter he produced in 1985. It is as up-to-date as we could make it; some changes being introduced at the page proof stage! Chapter 2, another update which tells how to use phage lambda to make genomic libraries, is testimony to the central role of this bacteriophage in DNA Cloning. The first author of Chapter 3 also contributed to the first edition. Here, however, we had more of a problem. The techniques for cDNA cloning have become more widely practised, not in the least because of the availability of cDNA cloning kits. While acknowledging the need to take full advantage of whatever is commercially available, Watson and Demmer describe techniques for cloning cDNA mainly in bacteriophage lambda vectors, but also indicate the practical value of PCR. Chapters 4 and 5 represent a departure from our previous format, and look at how to make and use probes both to screen for DNA clones and for analytical experiments: techniques that form the very basic armoury of a molecular biology laboratory. Dr Fritz, senior author of Chapter 6, was another contributor to the first edition. This area, *in vitro* mutagenesis, has become common place in the lab in the time since he wrote his original chapter. He has given a valuable description of what are now the most successful approaches for mutagenesis; a description that will greatly assist both those following his techniques directly, or making shortcuts with kits. The final chapter describes the basic techniques of DNA sequencing using the chain terminator approach. This is one of the single most powerful methods of analysing a cloned DNA. It is astounding how widespread DNA sequencing techniques have become, but there still is a need for a troubleshooting guide such as the one supplied here by Luke Alphey.

It is our hope that this book and its sister volumes will find their way onto bookshelves in laboratories, and in so doing will become messy and 'dog-eared'. It has been gratifying to see how widespread the use of the first edition has become, and a similar success would be rewarding to both the editors and the authors. Finally and most importantly we would like to thank all of the authors for their contributions and for their patience in bringing this project to fruition.

Dundee and Leeds   David M. Glover
September 1994   B. David Hames

Contents

Contents

Contents

5. The utilization of cloned DNAs to study gene organization and expression 143

Huw D. Parry and Luke Alphey

Contents

Contributors

LUKE ALPHEY
School of Biological Sciences, University of Manchester, Stopford Building, Oxford Road, Manchester M13 9PT, UK.

FREDRIC R. BLOOM
Life Technologies, Inc., Molecular Biology Research and Development, 8717 Grovemont Circle, Gaithersburg, MD 20884-9980, USA.

JEROME DEMMER
Biochemistry Department, University of Otago, Box 56, Dunedin, New Zealand.

HANS-JOACHIM FRITZ
Institut für Molekulare Genetik, Georg-August-Universität Göttingen, Grisebachstr. 8, D-37077 Göttingen, Germany.

DOUGLAS HANAHAN
Department of Biochemistry and Biophysics, Hormone Research Institute, University of California, San Francisco, CA 94143-0534, USA.

JOEL JESSEE
Life Technologies, Inc., Molecular Biology Research and Development, 8717 Grovemont Circle, Gaithersburg, MD 20884-9980, USA.

KIM KAISER
Department of Genetics, University of Glasgow, Church Street, Glasgow G11 5JS, UK.

HARALD KOLMAR
Institut für Molekulare Genetik, Georg-August-Universität Göttingen, Grisebachstr. 8, D-37077 Göttingen, Germany.

NOREEN E. MURRAY
Institute of Molecular and Cell Biology, University of Edinburgh, Darwin Building, King's Buildings, Mayfield Road, Edinburgh EH9 3JR, UK.

HUW D. PARRY
Department of Genetics, John Innes Institute, Norwich NR4 7UH, UK.

CHRISTINE J. WATSON
Roslin Institute (Edinburgh), Roslin, Midlothian EH25 9PS, Scotland, UK.

PAUL A. WHITTAKER
Department of Clinical Biochemistry, University of Southampton, Southampton General Hospital, Tremona Road, Southampton SO9 4XY, UK.

Abbreviations

BAC	bacterial artificial chromosome
BSA	bovine serum albumin
CCMB	calcium/manganese based transformation buffer
cDNA	complementary DNA
ccc DNA	covalently closed circular DNA
CTAB	cetyl trimethyl ammonium bromide
ddH$_2$O	deionized distilled water
ddNTP	2′,3′-dideoxynucleotide
DEPC	diethylpyrocarbonate
DH10B	strain of *E. coli*
DH5α	strain of *E. coli*
DMS	dimethyl sulfate
DMSO	dimethylsulfoxide
DTT	dithiothreitol
ds DNA	double-stranded DNA
EDTA	ethylenediamine tetraacetic acid
ES	embryonic stem
Exo V	exonuclease V
F$_c$	fraction of competent cells
FSB	frozen storage buffer
gd DNA	gapped duplex DNA
hd DNA	heteroduplex DNA
HPLC	high pressure liquid chromatography
IPTG	isopropyl-β-D-thiogalactopyranoside
JM109	strain of *E. coli*
MC1061	strain of *E. coli*
MHT	mix-heat-transform
M-MLV	Moloney murine leukaemia virus
mRNA	messenger RNA
PBS	phosphate-buffered saline
PCR	polymerase chain reaction
PCV	packed cell volume
PEG	polyethylene glycol
p.f.u.	plaque-forming units
PMSF	phenylmethylsulfonyl fluoride
PNK	polynucleotide kinase
RACE	rapid amplification of cDNA ends
RNase	ribonuclease
RT	reverse transcriptase

Abbreviations

RT-PCR	polymerase chain reaction coupled reverse transcription
SAM	S-adenosyl-L-methionine
SDS	sodium dodecyl sulfate
SSC	salt, sodium citrate
ss DNA	single-stranded DNA
SOB	bacterial growth medium
SOC	bacterial growth medium
TAE	Tris, acetic acid, EDTA
TB	bacterial growth medium
TBE	Tris, boric acid, EDTA
TCA	trichloroacetic acid
TE	Tris–HCl/EDTA
TEMED	N,N,N',N'-tetramethylethylenediamine
TES	Tris–HCl, NaCl, EDTA
TFB	standard transformation buffer
T_m	melting temperature
XFE	transformation efficiency
Xgal	5-bromo-4-chloro-3-indolyl-β-D-thiogalactopyranoside
YAC	yeast artificial chromosome

1

Techniques for transformation of *E. coli*

DOUGLAS HANAHAN, JOEL JESSEE, and FREDRIC R. BLOOM

1. Introduction

The transfer of plasmid DNA into *E. coli*, so-called 'plasmid transformation', is a cornerstone of modern molecular biology. The original observation that calcium ions induced a state of competence, whereby *E. coli* cells would take up DNA and establish it in the cell (1, 2), set the foundation for a number of techniques for inducing competence toward plasmid transformation. One class of techniques is founded on this original observation that Ca^{2+} and other divalent and multivalent cations can induce competence. More recently, an alternative technique, electroporation, involving application of a transient high voltage electrical field to a mixture of *E. coli* cells and plasmid DNA, has proved to be remarkably efficient. Electroshock transformation techniques currently produce the highest efficiencies of any known methodology. In this chapter we present a spectrum of techniques that together produce reliable transformations at differing efficiencies, and for distinctive purposes. These range from low efficiency colony transformations, which can be performed in a few minutes without advance preparation of competent cells, to electroshock transformations whereby high complexity plasmid libraries can be produced with minimal amounts of starting material. We also describe methods for properly storing *E. coli* strains that will be used for preparation of competent cells, as well as methods for producing frozen competent cells, and discuss other parameters which are important for reliable production of competent cells. This chapter is adapted from a more comprehensive presentation on methods for and genetic factors influencing plasmid transformation of *E. coli* and other bacteria (3).

2. Handling *E. coli* for competence induction and plasmid transformation procedures

Proper storage and handling of *E. coli* both prior to and subsequent of transformation procedures will improve efficiency, and especially increase reliability and reproducibility of the high efficiency methods. The protocols described below serve that purpose and are strongly recommended.

2.1 Storage and handling of *E. coli* before transformation

Despite the resilience of *E. coli*, the conditions under which cells are stored has been found to influence their subsequent ability to be rendered competent for transformation. *E. coli* maintained either in a suspension in glycerol at −20 °C or on agar plates at 4 °C both produce erratic competence when used to initiate cultures for preparing competent cells. Moreover, cells grown to saturation and nutrient limitation (e.g. the classical overnight) also produce erratic competence. Our recommendation is that the competence induction procedures described below be initiated from colonies on a freshly streaked agar plate derived from a frozen stock stored at −80 °C. An alternative is to use cells maintained in slow anaerobic growth in semi-solid agar stab bottles stored in the dark at room temperature (20–25 °C). These two storage conditions can supply reliable *E. coli* for competence induction, and protocols for each are presented below. When you first receive a strain of *E. coli* from another source it is especially important to prepare your own frozen stocks and/or stabs following *Protocols 1* and *2*. If possible you should also check for markers characteristic of the strain you are receiving. For example, if you received *E. coli* DH5α, the strain should:

(a) Not grow on antibiotic resistance plates (since it should not contain a plasmid).

(b) Be white (not blue) when grown on agar plates containing the fluorochromic lactose analogue Xgal (since it is *lacZ*− unless transformed with a plasmid carrying the α complementation unit).

(c) Grow in minimal medium containing glycerol as a carbon source (since lactose utilization is its only metabolic deficiency).

Protocol 1. Storing and retrieving *E. coli* in frozen stocks at −80 °C for competent cell preparations

Equipment and reagents

- SOB medium (see *Table 1*)
- SOB agar plates: autoclave SOB medium plus 1.5% agar—when cooled to 50 °C pour into 100 mm plastic Petri dishes (ideally in a sterile hood), and allow to cool
- Tungsten inoculating loop
- Snap-cap plastic tubes (e.g. Falcon 2059), or sterile metal-capped glass test-tubes
- Plating wheel (optional)
- Cryovials (e.g. Nunc Cat. No. 3–66656)

A. *Preparing frozen stocks of* E. coli

1. Use a flamed tungsten loop to collect some of the *E. coli* strain from an agar plate, a liquid culture, a previous frozen stock, or an agar stab bottle.

2. Spread the cells on an SOB agar-filled Petri dish (the plate) by: streak-

---

2

1: Techniques for transformation of E. coli

ing the loop of cells across the agar surface, flaming the loop, and re-streaking with partial overlap from the previous streak, repeated again. Alternatively make a radial streak with the loop of cells, flame the loop, and then place the loop gently on the plate and move it radially inward as the plate is spun on a plating wheel.

3. Incubate the plate at 30–37 °C until 2–3 mm diameter colonies form (typically overnight). Pick several colonies with a flame sterilized tungsten loop and inoculate 5–10 ml of SOB medium in a sterile snap-cap plastic tube, or a metal-capped glass test-tube.

4. Incubate the tube at 37 °C with agitation until the culture is midway into the logarithmic phase of growth. The cell density should be $1-5 \times 10^8$ cells/ml; which is typically an $OD_{550} = 0.5 - 0.7$; at this density the cells are visible as a cloudy suspension.

5. Dilute the culture 1:1 with a solution that is 60% SOB medium/40% glycerol, to give a final concentration of 20% glycerol. Mix thoroughly by gentle vortexing.

6. Aliquot 0.5–1 ml of the cell suspension into a numbered series of 2 ml cryovials. Chill the tubes on ice, and flash freeze them by submerging only the bottom halves in a freshly prepared dry ice/ethanol bath. Place the tubes at -70 to -80 °C.

B. *Retrieving* E. coli *from* -80 °C *frozen stocks*

1. Remove a tube of the frozen stock from the freezer and place on dry ice.

2. Quickly scrape a clump of frozen cells from the surface of the frozen suspension, using for example a sterile yellow micropipette tip, or a flamed tungsten loop. Replace the frozen stock in the freezer.

3. Touch the frozen clump of cells (now thawing) on to an SOB agar plate, where it will immediately thaw.

4. Spread the puddle of medium to disperse the cells as described in part A, step 2, and then incubate the plate at 30–37 °C to establish colonies. Inspect the colonies to confirm typical *E. coli* colour and morphology.

Each tube of frozen cells can be repeatedly used as a source of cells. However, the warming during removal from the freezer will reduce cell viability over time. Thus we recommend that a serially numbered set of tubes be prepared, and each tube used 10–20 times. However, if the titre of viable cells in a clump declines precipitously (i.e. only a few colonies develop) before this number of retrievals is reached, the tube should be discarded and the next aliquot used in its place. The storage lifetime of *E. coli* in this condition is essentially indefinite, at least five to ten years, provided that the cells do not experience repeated freeze/thaw cycles.

Protocol 2. Storage of *E. coli* in semi-soft agar stab bottles

Equipment and reagents

- Bijou bottles—made by United Glass, Stock Number 528, and available from Johns Scientific, 175 Hanson Street, Ontario, M4C 1A7, Canada
- SOB medium plus 0.8% Bacto agar (see *Table 1* for SOB formulation)
- Tungsten inoculating loop

A. *Making the stab bottles*

1. Prepare a solution of SOB medium plus 0.8% Bacto agar. The components can be added, and the suspension heated briefly in a microwave oven to dissolve the agar.

2. Aliquot 4 ml of the hot medium into 5 ml glass Bijou bottles (or other screw-cap bottles that can be tightly sealed). Place the caps loosely on the bottles. Autoclave to sterilize.

3. Leave the bottles loosely capped both during cooling, and overnight at room temperature, to allow evaporation of condensed water. Then seal tightly and store in the dark at room temperature.

B. *Preparing* E. coli *stab cultures*

1. First streak the *E. coli* to be stored on an agar plate (see *Protocol 1*) and incubate to develop fresh 2–3 mm diameter colonies. Inspect to confirm typical *E. coli* colour and morphology.

2. Pick up several well isolated colonies with a flame sterilized tungsten inoculating loop.

3. Briefly flame the cap of the stab bottle to kill surface bacteria. Remove the cap, and stab the loop of cells into the agar. Quickly remove the loop and replace the cap, screwing it tightly to seal the bottle.

4. Store the 'stab bottle' in the dark at room temperature. **Do not store stabs of *E. coli* at 4 °C.** This temperature reduces long-term viability and affects the reliability of transformations originating from cells stored in such conditions.

C. *Retrieving* E. coli *from stab cultures*

1. Inspect the stab bottle, which should show opaque regions emanating from the original stab marks, but not extreme discoloration or surface growth, which could be indicative of bacterial contamination.

2. Flame the stab bottle cap briefly, and flame sterilize a tungsten inoculating loop.

3. Remove the cap, stab the loop into the agar in one or two places, remove the loop, and replace the cap.

4

4. Streak the loop of cells on a SOB agar plate to disperse the cells (*Protocol 1A*), and incubate at 37 °C to establish colonies.

5. Inspect the colonies for typical *E. coli* colour (opaque white) and morphology, as contaminating bacteria can overgrow an *E. coli* stab culture and be mistakenly employed in subsequent experiments.

E. coli stored in the dark at room temperature in Bijou bottles with 4 ml of medium have a considerable lifetime, up to five years for many strains. When used for producing competent cells, we advise preparation of fresh stabs on a yearly basis. Serial passaging from stab to stab is not recommended. Rather, passages should be separated by streaking from the old stab on to an agar plate to establish isolated colonies, which are then used to inoculate the fresh stab. This practice prevents bacterial contaminants from overgrowing the culture. For example, *Clostridium* has been known to overgrow *E. coli* cultures and be mistakenly employed for preparation of competent cells. Although stab cultures can be a reliable source of *E. coli* for transformation procedures, it is nevertheless our strong recommendation that frozen stocks be used whenever possible as a source for initiating competent cell procedures.

2.2 Dispersing transformed cells on agar plates to establish colonies (plating)

E. coli cells are somewhat fragile following competence induction and DNA transformation procedures, which necessitates gentle treatment following the heat shock or electroshock, the subsequent addition of medium, and the 37 °C 'recovery' period. In particular, the vigorous spreading of cells on a dry agar plate can significantly reduce the number of transformed cells that establish colonies following a transformation. Therefore the following procedure is recommended.

Protocol 3. Plating transformed *E. coli* cells

Equipment and reagents

- SOB agar plates with appropriate antibiotic—prepare as in *Protocol 1*, adding antibiotic when the autoclaved medium has cooled to 50 °C. For plasmids carrying the β-lactamase antibiotic resistance gene, add 70 μg/ml carbenicillin, or 100 μg/ml ampicillin. The plates should be incubated at room temperature until surface condensation evaporates, then stored at 4 °C, and used within two weeks

- A plating wheel (optional)
- A glass spreading tool. Take a 9 cm Pasteur pipette, flame seal the tip, then make a 90° bend midway across the narrow section, producing an L-shape. The tip can be bent slightly in the perpendicular plane to match the meniscus at the edge of the agar plate

A. *Method*

1. Spot 150–200 μl of growth medium (e.g. SOB) in the middle of a 10 cm SOB agar plate containing the appropriate antibiotic for the plasmid's

---

Protocol 3. *Continued*

drug resistance gene. Aliquot the desired volume of the transformed cell suspension into the middle of the puddle.

2. Quickly spread the puddle evenly over the plate using an L-shaped glass spreading tool. Your spreading should be completed **before** the medium dries into the plate, which can be visualized by an even surface reflection off a ceiling light (or hood light if plating in a biosafety hood). If the medium dries into the plate before spreading is complete, then make a larger puddle of medium when preparing subsequent plates. A plating wheel is very useful for evenly spreading the cell suspension.

3. Leave the plates at room temperature for 10–30 min to allow the cell suspension to dry. Then incubate the plates at 30–37 °C to develop colonies of transformed cells.

4. If an entire 1 ml transformation is to be spread on a single 10 cm plate, the cells should first be concentrated by gentle centrifugation (5 min at 600–800 g, or 1500–2000 r.p.m. in a table-top centrifuge). After decanting the supernatant, gently resuspend the pellet of cells with 250 μl of SOC medium (*Table 1*), and plate as described above, omitting the initial puddle of medium.

2.3 Storage of transformed cells

Chemically competent cells in their mixture of transformation buffer (20%) and SOC medium (80%) can be stored overnight at 4 °C, but with some loss of viability, which varies with strain and transformation procedure, and can be considerable (> 90% loss). We suggest that the transformed cells be washed to remove the transformation chemicals, as described in *Protocol 4*, for storage overnight. *E. coli* transformed by electroshock technique (*Protocol 9*) do not require this washing procedure for short-term storage. For long-term storage (more than several days), all transformed *E. coli* should be frozen and maintained at −80 °C. *Protocol 4* describes methods for short- and long-term storage of chemically competent transformed cells.

Note: for electrocompetent cells, adjust the suspension of transformed cells in SOC medium to 20% in glycerol by adding an equal volume of a sterile solution of 40% glycerol plus 60% SOB medium, mix thoroughly, and proceed with *Protocol 4*, step 4.

Protocol 4. Storage of chemically transformed cells

Equipment and reagents

- SOC medium (*Table 1*), filter sterilized
- SOB medium (*Table 1*) plus 20% glycerol, filter sterilized
- 1.5–2 ml polypropylene tubes (e.g. 1.5 ml microcentrifuge snap-cap or 2 ml screw-cap cryotubes with polypropylene seals)

Method

1. Pellet the transformed cells in a clinical centrifuge (5 min, 1500–2000 r.p.m., 600–800 *g*). Decant the supernatant carefully, so as to avoid dislodging the cell pellet.

2. For short-term storage (one to two days), resuspend the cells in 0.5–1 ml SOC medium by gently vortexing or pipetting. Place the cells at 4 °C for up to several days (the viability of the transformed cells should remain > 90%).

3. For long-term storage, resuspend the cell pellet in a mixture of 80% SOB medium/20% re-distilled glycerol.

4. Transfer the suspension of transformed cells (in volumes ranging from 100 μl to the entire contents) into 1.5–2 ml polypropylene tubes, preferably screw-capped. Chill on ice, flash freeze by partially submerging the tubes in a freshly prepared dry ice/ethanol bath, and place at −80 °C.

2.4 Reagent quality

Both chemical and electroshock methods of transformation show some sensitivity to the quality of reagents used in the transformation buffers. As a general rule ultrapure or reagent grade chemicals should be employed, and most major suppliers are satisfactory. The quality of the water is very important for the procedures using DMSO, and can impact upon all of the methods. In general we recommend the use of water purified by reverse osmosis, as it effectively removes not only salts but also organic contaminants, including those with similar volatility to water, which are not removed by distillation. However, it is important that the carbon filters which remove organics be replaced regularly, as there is no gauge to measure their saturation (except drastically reduced competence when it occurs). Distilled or deionized water can also be further purified in the laboratory with carbon, as described previously (4).

Dimethylsulfoxide (DMSO) is subject to oxidation, and its oxidation products can be a major source of variability in the levels of competence induced by procedures which employ it. Note also that DMSO dissolves polystyrene, and therefore tubes composed of this material cannot be used. We recommend that DMSO be purchased in the smallest possible amounts (e.g. 100 ml bottles) at spectroscopy grade, aliquoted to fill small (0.5 ml) polypropylene tubes completely, which are then stored at −80 °C. A tube of DMSO is thawed, used for that day, and then discarded. Similarly, the dithiothreitol (DTT) used in the TFB method (*Protocol 6*) is also subject to impurities which inhibit competence. One of the most reliable suppliers of DTT is Calbiochem. Impurities in glycerol products can affect all of the competence and storage protocols used here, and it is recommended that re-distilled or spectroscopy grades of glycerol be used. Among the recommended suppliers are Fluka and GIBCO-BRL.

Other variables which can influence competence induction are the flasks and tubes used for culturing and treating the cells during transformation. Of particular concern is soap residues in sterilized glassware. You should dedicate a set of flasks and other reusable containers (e.g. centrifuge tubes for concentrating electrocompetent cells). When first acquired for transformation procedures, the flasks should be cleaned thoroughly to remove all soap and other material. For glass shake flasks, first acid wash, then ethanol wash, and finally rinse extensively with ultrapure water.

For routine use in preparation of competent cells, the dedicated glassware (and other reusable tubes and flasks) used in transformation procedures should be thoroughly rinsed with distilled water after each use, and then autoclaved three quarters full of water, to prevent deposition of surface residues. For *E. coli* it is not necessary to use soap or glassware washing services; rinsing the flasks by hand in water without soap, followed by steam sterilization, will suffice.

A second source of transformation inhibitors comes from surfactants used in manufacturing certain brands of polypropylene tubes. Among the reliable suppliers of polypropylene tubes for transformation procedures are Falcon and Corning.

The components of the growth media can also influence transformation. We routinely use Bacto tryptone and Bacto yeast extract. Alternative suppliers should be carefully compared, as lot variations have been reported with regard to transformation experiments. Again reagent grade chemicals and the purest available water are recommended.

2.5 Formulations of media used in transformation procedures

All of the transformation protocols described herein use a common set of growth media, with the only major distinction being in the presence or absence of elevated levels of magnesium. These differ from many traditional media in that the levels of sodium are low. The recovery of cells from all of these procedures is enhanced in SOC medium, which contains glucose. We recommend that these media be filter sterilized just prior to use, to remove particulate matter and any slow growing contaminating micro-organisms. Each media is identical to or derived from a stock of SOB−Mg medium, and their formulations are described in *Table 1*.

Although SOB media can be used for subsequent growth of transformed *E. coli*, we recommend another medium, TB, for large scale expansion of transformed clones (5). TB medium (*Table 2*) has proved to be particularly beneficial for maximizing the yield of plasmid DNA from *E. coli* transformed with plasmids based on the high copy number pUC vectors. In addition, this

Table 1. Formulations of SOB based media used in transformation procedures[a]

Compound	Amount/litre	Final concentration
Growth medium SOB−Mg		
Bacto tryptone	20 g	2.0%
Bacto yeast extract	5 g	0.5%
NaCl	10 ml of a 1 M stock	10.0 mM
KCl	2.5 ml of a 1 M stock	2.5 mM
Growth medium SOB		
SOB−Mg medium	1 litre	99%
$MgCl_2$ + $MgSO_4$	10 ml of a 2 M stock	20 mM
Recovery medium SOC		
SOB−Mg medium	1 litre	98%
$MgCl_2$ + $MgSO_4$	10 ml of a 2 M stock	20 mM
Glucose	10 ml of a 2 M stock	20 mM

[a] To prepare SOB−Mg based media, first combine tryptone, yeast extract, NaCl, and KCl in purest available water. Aliquot into pre-rinsed glass flasks, and autoclave for 30–40 min to generate SOB−Mg medium. Sterile filter just prior to use for protocols requiring this medium itself. Prepare a 2 M stock of Mg^{2+} by combining 203 g/litre $MgCl_2.6H_2O$ and 247 g/litre $MgSO_4.7H_2O$ in purest water, and sterilize by filtration through a pre-rinsed 0.2 μm filter unit. Similarly prepare a 2 M stock of glucose (360 g/litre with purest water), sterile filter, and store frozen in aliquots. For SOB and SOC, combine these reagents as specified with SOB−Mg medium, then sterile filter through a 0.2 μm filter unit just prior to use. It is advisable to use detergent-free filter units for all of these purposes. The final pH of these media should be 6.8–7.2.

Table 2. Formulation of TB medium for plasmid DNA production

Compound	Amount/litre	Final concentration
Media component		
Bacto tryptone	12 g	1.2%
Bacto yeast extract	24 g	2.4%
Re-distilled glycerol	4 g	0.4%
Combine in 900 ml H_2O and autoclave		
Phosphate component		
KH_2PO_4	2.3 g	17 mM
K_2HPO_4	12.5 g	72 nM

Dissolve the two potassium phosphates in 100 ml H_2O and autoclave. When the media and phosphate solutions are cool, combine them to give 1 litre of TB.

medium, when used in conjunction with low temperature growth conditions (30 °C), dramatically improves the amplification of plasmids that carry unfavourable sequences that are prone to rearrangement (3) (J. Jessee, unpublished observations).

3. Choice of transformation method

Bacterial transformations can be divided into three classes with regard to purpose and consequent requirements for transformation efficiency.

(a) Primary cloning of high complexity plasmid populations, for example cDNAs representing an mRNA population, or fragments of an organism's genome.

(b) Cloning of more restricted populations of plasmids, for example following *in vitro* manipulations such as oligonucleotide mutagenesis or subcloning fragments of previously cloned DNAs.

(c) The re-introduction of a pure plasmid population (a single clone) for purposes such as the preparation of large quantities of plasmid DNA.

Table 3 summarizes the recommended protocols for these various purposes.

Re-introduction of cloned plasmid DNA into *E. coli* does not require cells with high competence, and various methods will suffice. Among these are the rapid colony transformation (*Protocol 8*), or frozen competent cells (*Protocols 5* and *7*). If frozen competent cells are used, we suggest older lots that have decayed in competence, since these are nevertheless more than adequate for the purpose. Alternatively, a small clump of highly competent frozen cells can be quickly scraped off the surface of a frozen aliquot, which is returned immediately to the freezer without thawing. A 5–10 μl clump of high competence cells is sufficient for retransforming a cloned plasmid into *E. coli*.

The intermediate purpose of manipulating and modifying previously cloned DNA usually requires transformation efficiencies in the range of 10^7–10^8 transformants/μg. These values are achieved with the immediate and frozen storage versions of the high efficiency transformation methods (*Protocols 5–7*). Experience will to some extent dictate the choice of method. If insufficient numbers of transformants are being generated, the options are to increase either the number of discrete transformations performed, or the amount of DNA being used, or change the type of transformation procedure (to improve the competence). For some types of *in vitro* modifications, very high efficiencies may be necessary to obtain the desired number of transformants, in which case the highest efficiency methods should be used.

Plasmid transformations that are intended to produce large numbers of clones representing high complexity populations, or are limited by very small amounts of available plasmid DNA, require very high efficiency transformation protocols. It is increasingly clear that electroshock transformation is the most efficient, and therefore is the method of choice for such purposes. The procedure suffers from the requirement for a high voltage electroporation device, and from the large volume of cells that must be grown during the preparation of electrocompetent cells. This latter problem is being obviated

Table 3. *E. coli* transformation methods

Protocol	Method	Typical transformation efficiency (XFE)[a]	Typical competence (F_c)[b]	Best for strains	Recommended applications
5	Calcium manganese based (CCMB)	$1–10 \times 10^8$	1–5%	MC1061, DH10B, DH12S	cDNA cloning, mutagenesis, high efficiency cloning
6	TFB based high efficiency	$0.4–2 \times 10^9$	2–10%	DH5α, HB101, DH11S	cDNA cloning, mutagenesis, high efficiency cloning
7	FSB based frozen competent	$1–5 \times 10^8$	1–4%	DH5α, HB101, DH11S	cDNA cloning, mutagenesis, high efficiency cloning
8	Rapid colony transformation	$10^3–10^4$	<0.001%	Most common cloning hosts	Re-introduction of cloned plasmids into *E. coli*
9	Electroshock	$0.5–5 \times 10^{10}$	80%	Most strains, especially MC1061, DH10B, DH12S	High complexity cDNA and genomic cloning

[a] The transformation efficiency (XFE) is given here in transformants (colonies formed) per microgram of the plasmid pBR322.
[b] F_c is the fraction of viable cells that is competent for plasmid transformation. See Section 7 for discussion of these parameters.

by the availability of commercial preparations of concentrated cells for electroshock transformation.

An alternative to electroshock transformation is the use of the high efficiency chemical transformation procedures under conditions that produce optimal competence. Both *Protocol 7* with DH5α and *Protocol 5* with DH10B will produce efficiencies in excess of 10^9 transformants/μg pBR322, provided that all the important parameters of the protocols are carefully satisfied. Given careful attention to the transformation protocols, it is possible to routinely achieve high efficiency competent cell preparations. It is noted in addition that frozen competent cells in all four competence ranges are commercially available, including maximally efficient chemically competent cells, and electrocompetent cells. This alternative may prove reasonable for investigators who prefer to avoid the rigours of achieving routine high efficiency transformations. However, all of the procedures presented herein will provide reliable transformations, given commitment to the details of the protocols.

4. Protocols for preparing competent *E. coli* using chemical treatments

We present two alternative chemical treatment regimens for inducing high degrees of competence. Each seems to be favoured by different strains of *E. coli*; it is likely that genetic differences in the cell envelopes determine susceptibility to transformation by one or the other of these conditions (J. Jessee and F. Bloom, unpublished results). The first method extends upon the original Ca^{2+} based procedure through the additional utilization of manganese and potassium. The second method originated from a series of experiments into the mechanisms of transformation (6). This protocol produces very high transformation efficiencies using DH5α and many other strains, and is considered complex in its conditions, in that dimethylsulfoxide (DMSO), dithiothreitol (DTT), and hexamine cobalt trichloride are employed in addition to the cations used in the simpler Mn^{2+}/Ca^{2+} based version. Each method is suitable for either immediate transformation or for frozen storage of competent cells. In addition, a rapid transformation protocol for re-introduction of cloned plasmids into *E. coli* is presented.

4.1 Moderate efficiency methods based upon Mn^{2+}/Ca^{2+} treatment

MC1061 (7) and its derivative, DH10B (8), are preferentially transformed by a modification of the Mandel and Higa method (1). These strains differ from those derived from Hoffman Berling strain 1100 (e.g. DH5, DH5α, DH5αMCR) as well as many other strains (e.g. HB101, C600) in that Mg^{2+} is **not** beneficial in the growth medium, and the addition of either DMSO or

DTT to the transformation buffer **reduces** competence. The following protocol works very well for MC1061 derivatives and may be preferable for certain other *E. coli* strains as well.

Protocol 5. Calcium/manganese based (CCMB) transformation of MC1061, DH10B, and related strains of *E. coli*

Equipment and reagents

- Plate of fresh *E. coli* colonies (*Protocol 2*)
- SOB−Mg medium (*Table 1*)
- Dedicated glass shake flasks, sterilized—use metal-capped Erlenmeyer flasks (Belco Cat. No. 2510−00300) or cotton plugged Fernbach flasks (Belco Cat. No. 2551−02800); (see note in Section 2.4 regarding cleaning)
- 50 ml screw-cap polypropylene tubes (e.g. Falcon 2070)
- CCMB transformation buffer (*Table 4*)
- 10 ml snap-cap polypropylene tubes (e.g. Falcon 2059)
- Cryovials with polypropylene seals (e.g. Nunc Cat. No. 3−66656)
- SOC medium (*Table 1*)
- TE: 10 mM Tris−HCl Ph 8.0, 1 mM EDTA

Method

1. Pick several 2−3 mm diameter colonies off a freshly streaked plate into 1 ml SOB−Mg growth medium (SOB without Mg). Avoid collecting agar fragments. Vortex gently to disperse cells.
2. Inoculate the dispersed colonies into a shake flask containing SOB−Mg medium (use 50 ml medium in a 500 ml non-baffled Erlenmeyer flask, or 200 ml in a 2.8 litre Fernbach flask).
3. Incubate at 275 r.p.m., 37 °C until the OD_{550} reaches 0.3, which corresponds to 5×10^7 cells/ml.
4. Collect the cell suspension into sterile 50 ml polypropylene centrifuge tubes and chill on ice for 10 min. (Take a 10 µl aliquot of cells to determine the density of viable cells by plating a 10^6 dilution on to a SOB agar plate.)
5. Pellet the cells at 750−1000 *g* (2000−3000 r.p.m. in a clinical centrifuge) for 10−15 min at 4 °C. Decant the supernatant and invert the tubes to remove excess culture medium.
6. Disperse the cells in 1/3 vol. of CCMB by gentle vortexing, or rapping of centrifuge tube.
7. Incubate on ice for 20 min.
8. Centrifuge at 750−1000 *g* for 10 min. Decant as in step **5**.
9. Resuspend the cells in CCMB at 1/12 of original volume.
10. The competent cells may then be:
    (a) **Used immediately**, following steps **11−14**.
    (b) **Stored frozen** for subsequent use, by aliquoting 0.4−1.0 ml into chilled 2 ml cryovials, which are flash frozen by **partial submersion**

13

Protocol 5. *Continued*

in a freshly prepared dry ice/ethanol bath, and stored at −80 °C. To use the frozen competent cells, remove a tube(s) from the freezer, thaw on ice, and follow steps **11–14** below.

11. Aliquot 200 μl of the competent cell suspension into chilled Falcon 2059 polypropylene tubes. Incubate on ice for 10 min.

12. Add the DNA in less than 10 μl of TE, or 1–2 μl of an undiluted ligation reaction. Incubate on ice for 30 min.

13. Heat shock in a water-bath at 42 °C for 90 sec. Place on ice for 2 min.

14. Add 800 μl SOC medium. Incubate at 37 °C with mild agitation for 60 min. Then plate on to SOB agar plates with appropriate drug selection (*Protocol 3*) and incubate at 37 °C to develop colonies of the transformed cells.

Table 4. Formulation of transformation buffer CCMB

Compound	Amount/litre	Final concentration
Potassium acetate	10 ml of a 1 M stock (pH 7.0)	10 mM
Re-distilled glycerol	100 g	10% (w/v)
$CaCl_2.2H_2O$	11.8 g	80 mM
$MnCl_2.4H_2O$	4.0 g	20 mM
$MgCl_2.6H_2O$	2.0 g	10 mM

First prepare a 1 M solution of potassium acetate, and adjust to pH 7.0 using KOH, then sterilize by filtration through a pre-rinsed 0.22 μm membrane, and store frozen. Then prepare a solution of 10% potassium acetate, 10% glycerol using these reagents and the purest available water. Add salts sequentially as solids, allowing each to enter into solution before adding the next. Adjust pH to 6.4 with 0.1 M HCl. Do not adjust pH upward with base. Sterilize the solution by filtration through a pre-rinsed 0.22 μm filter, and store at 4 °C. CCMB is stable indefinitely.

4.2 High efficiency transformation methods utilizing complex conditions (TFB/FSB)

The high efficiency transformation procedures described below are effective with many *E. coli* strains, including several which have wide applicability in molecular cloning experiments (DH5α, HB101). The development of this method and its evaluation has been described in considerable depth previously (3, 4, 6). Critical evaluation, as well as troubleshooting strategies, important parameters, genetic influences, and variations (e.g. for the strain χ1776) can be found in these papers. The key parameters for success with this method are:

● the growth of the cells in medium containing magnesium (20 mM)
● collection of the cells during midlog phase growth following an 'appropriate' inoculation regimen

14

- the use of ultrapure water lacking organic contaminants
- using DMSO free of significant oxidation

When this method is used with these parameters satisfied, very high transformation frequencies can routinely be achieved (2–3% of the plasmid molecules affecting a transformed cell; 5–10% of the cells rendered competent for transformation). On average, the immediate transformation procedure (*Protocol 6*) gives higher frequencies than does the frozen competent version (*Protocol 7*), which of necessity omits DTT. Interestingly, some commercial preparations of competent cells often equal or exceed those obtained with these protocols, and can be considered as a valid option to preparing high efficiency competent cells in the laboratory.

Protocol 6. TFB based chemical transformation protocol

Equipment and reagents

- Plate of fresh *E. coli* colonies (*Protocol 2*)
- SOB medium (*Table 1*)
- Sterile shake flasks (*Protocol 5*)
- 50 ml screw-cap polypropylene tubes (Falcon 2070)
- TFB transformation buffer (*Table 5*)
- 12 ml snap-cap polypropylene tubes (Falcon 2059)

- DnD solution: to prepare, combine 9 ml of spectroscopy grade DMSO (Fluka Cat. No. 1641), with 1.53 g DTT (ultrapure, Calbiochem), and 100 μl of a 1 M sterile filtered stock of potassium acetate, pH 7.5; aliquot 0.25 μl into 0.5 ml microcentrifuge tubes and store frozen, use each tube once and discard
- SOC medium (*Table 1*)

Method

1. Pick several 2–3 mm diameter colonies off a freshly streaked SOB agar plate and disperse in 1 ml SOB medium by vortexing. Use one colony per 10 ml of culture medium. The cells are best streaked from a frozen stock or fresh stab about 16–20 h prior to initiating liquid growth.

2. Inoculate the cells into an Erlenmeyer flask containing SOB medium. Use a culture volume to flask volume of between 1:10 and 1:30 (e.g. 30–100 ml in a 1 litre flask).

3. Incubate at 37 °C with moderate agitation until the cell density is 4–7 × 10^7 viable cells/ml (OD$_{550}$ = 0.5 for DH5α).

4. Collect the culture into 50 ml polypropylene centrifuge tubes (such as Falcon 2070 tubes) and chill on ice for 10–15 min. (Take a 10 μl aliquot of cells to determine the density of viable cells by plating a 10^6 dilution on to a SOB agar plate.)

5. Pellet the cells by centrifugation at 750–1000 *g* (2000–3000 r.p.m. in a clinical centrifuge) for 12–15 min at 4 °C. Drain the pelleted cells thoroughly, by inverting the tubes on paper towels and rapping gently to remove any liquid. A micropipette can be used to draw off recalcitrant drops.

Protocol 6. *Continued*

6. Resuspend the cells in 1/3 of the culture volume of TFB by vortexing moderately. Incubate on ice for 10–15 min.

7. Pellet the cells and drain thoroughly as in step **5**.

8. Resuspend the cells in TFB to 1/12.5 of the original volume. This represents a concentration from each 2.5 ml of culture into 200 μl of TFB.

9. Add DMSO and DTT solution (DnD) to 3.5% (v/v) (7 μl per 200 μl of cell suspension). Squirt the DnD into the centre of the cell suspension and immediately swirl the tube for several seconds. Incubate the tubes on ice for 10 min.

10. Add a second equal aliquot of DnD as in step **9** to give a 7% final concentration. Incubate the tubes on ice for 10–20 min.

11. Pipette 210 μl aliquots into chilled 17 × 100 mm polypropylene tubes (Falcon 2059 or their equivalent).

12. Add the DNA solution in a volume of < 10 μl, swirling to mix. Incubate the tubes on ice for 20–40 min. (Ligation reactions should be diluted at least 5 ×, or precipitated in ethanol, and resuspended in 0.5 × TE.)

13. Heat shock the cells by placing tubes in a 42 °C water-bath for 90 sec. Return the tubes into ice to quench the heat shock, allowing 2 min for cooling.

14. Add 800 μl of SOC medium to each tube. Incubate at 37 °C with moderate agitation for 30–60 min. (This incubation period should be omitted for M13 transfections.)

15. Spread the cells on agar plates (*Protocol 3*) containing appropriate antibiotics (or other conditions), and incubate at 37 °C to select for transformants.

Table 5. Standard transformation buffer (TFB)

Compound	Amount/litre	Final concentration
K-MES	20 ml of 0.5 M stock (pH 6.3)	10 mM
KCl (ultrapure)	7.4 g	100 mM
$MnCl_2.4H_2O$	8.9 g	45 mM
$CaCl_2.2H_2O$	1.5 g	10 mM
$HACoCl_3$	0.8 g	3 mM

(Final pH 6.2 ± 0.10)

Equilibrate a 0.5 M solution of MES (2[*N*-morpholino]ethone sulfonic acid) to pH 6.3 using concentrated KOH, then sterilize by filtration through a 0.22 μm membrane, and store in aliquots at −20 °C. Make a solution of 10 mM K-MES, using the 0.5 M MES stock and the purest available water. Add salts sequentially as solids, allowing each to enter into solution before adding the next. Then filter the solution through a 0.22 μm pre-rinsed membrane. Aliquot into sterile flasks, and store at 4 °C. TFB is stable for more than one year.

The preparation of frozen competent cells using this method requires a different transformation buffer (FSB), and the omission of DTT.

Protocol 7. Frozen storage of competent cells using FSB
conditions

Equipment and reagents

- Fresh plate of *E. coli* colonies (*Protocol 2*)
- SOB medium (*Table 1*)
- Sterile shake flask (*Protocol 5*)
- 50 ml screw-cap polypropylene tubes (e.g. Falcon 2070)
- FSB transformation buffer (*Table 6*)
- 12 ml snap-cap polypropylene tubes (e.g. Falcon 2059)

- DMSO: use spectrophotomeric grade (e.g. Fluka Cat. No. 41641); aliquot to fill 0.5 ml microcentrifuge tubes and store frozen; use each tube once and discard
- Cryovials with polypropylene seals (e.g. Nunc Cat. No. 3–66656)
- SOC medium (*Table 1*)

A. *Preparation*

1. Pick several 2–3 mm diameter colonies off a freshly streaked SOB agar plate and disperse in 1 ml SOB medium by vortexing. Use one colony per 10 ml of culture medium. The cells are best streaked from a frozen stock or a fresh stab about 16–20 h prior to initiating liquid growth.

2. Inoculate the cells into an Erlenmeyer flask containing SOB medium. Use a culture volume to flask volume of between 1 : 10 and 1 : 30 (e.g. 30–100 ml in a 1 litre flask).

3. Incubate at 37 °C with moderate agitation until the cell density is 6–9 × 10^7 viable cells/ml ($OD_{550} = 0.5$ for DH5α).

4. Collect the culture into 50 ml polypropylene centrifuge tubes (Falcon 2070 tubes) and chill on ice for 10–15 min. (Take an aliquot of cells to determine viable cell density by plating a 10^6 dilution on to a SOB agar plate.)

5. Pellet the cells by centrifugation at 750–1000 *g* (2000–3000 r.p.m. in a clinical centrifuge) for 12–15 min at 4 °C. Drain the pelleted cells thoroughly, by inverting the tubes on paper towels, and rapping gently to remove any liquid. A micropipette can be used to draw off recalcitrant drops.

6. Resuspend the cells in 1/3 of the culture volume of frozen storage buffer (FSB) by vortexing moderately. Incubate on ice for 10–15 min.

7. Pellet the cells as before, and drain thoroughly as in step **5**.

8. Resuspend the cells in FSB to 1/12.5 of the original volume. (Each 2.5 ml of culture is concentrated into 200 µl of FSB.)

9. Add DMSO to 3.5% (v/v) (7 µl per 200 µl of cell suspension). Squirt the DMSO into the centre of the cell suspension and immediately swirl

---

17

Protocol 7. *Continued*

the tube for several seconds. Incubate the tube on ice for 5 min. (DTT is not used in this protocol.)

10. Add a second equal aliquot of DMSO, as before, giving a 7% final concentration. Incubate on ice for 10–15 min.

11. Pipette 210 μl aliquots into chilled 2 ml cryovials.

12. Flash freeze by **partially submerging** the cryovials upright in a rack in a freshly prepared dry ice/ethanol bath. (Be careful that alcohol does not get inside the tubes. Avoid total immersion but instead just set the bottom half of the tubes into the bath. Use fresh 95% ethanol to prepare the bath.)

13. Transfer the tubes to a −80 °C freezer.

B. *Use of frozen competent cells*

1. Remove the tube(s) from the freezer and thaw on ice. If small vials were used, it is recommended that the cells be transferred in 200 μl aliquots into 12 ml polypropylene tubes (e.g. Falcon 2059).

2. Add the DNA solution in a volume of <10 μl. Swirl the tube to mix the DNA evenly with the cells. (Ligation reactions should be diluted at least 5 ×, or precipitated in ethanol, and resuspended in 0.5 × TE.)

3. Incubate the tube(s) on ice for 10–30 min.

4. Heat shock the cells by placing the tubes in a 42 °C water-bath for 90 sec, and then chill by returning the tubes immediately to 0 °C (crushed ice).

5. Add 800 μl of SOC medium and incubate at 37 °C with moderate agitation for 30–60 min. (This incubation period should be omitted for M13 transfections.)

6. Spread the cells on agar plates (*Protocol 3*) containing appropriate antibiotics (or other conditions), and incubate at 37 °C to select for transformants.

4.3 Rapid transformation using colonies on agar

It is possible to collect a few colonies of cells from an agar plate, disperse them in transformation buffer, chill on ice to induce competence, and add DNA to effect transformation. This procedure is extremely fast and easy. It is limited in efficiency, typically giving 10^4 transformants/μg pBR322. It is not recommended for primary cloning experiments, but is very useful for reintroducing cloned plasmid DNA back into *E. coli* prior to growing up large scale cultures for plasmid DNA purification and other purposes.

Table 6. Frozen storage buffer (FSB)

Compound	Amount/litre	Final concentration
Potassium acetate	10 ml of a 1 M stock (pH 7.5)	10 mM
Re-distilled glycerol	100 g	10% (w/v)
KCl	7.4 g	100 mM
$MnCl_2.4H_2O$	8.9 g	45 mM
$CaCl_2.2H_2O$	1.5 g	10 mM
$HACoCl_3$	0.8 g	3 mM

(Final pH 6.20 $\pm$ 0.10)

Equilibrate a 1 M solution of potassium acetate to pH 7.5 using KOH, then sterilize by filtration through a 0.22 μm membrane and store frozen. Prepare a 10 mM potassium acetate, 10% glycerol solution using this stock and the purest available water. Add salts sequentially as solids, allowing each to enter into solution before adding the next. Adjust the pH (if necessary) to 6.4 using 0.1 M HCl. Do not adjust the pH upward with base. The pH may drift for one or two days before settling at 6.1–6.2. Sterilize the solution by filtration through a pre-rinsed 0.22 μm membrane and store at 4°C. $HACoCl_3$ is hexamine cobalt trichloride.

Protocol 8. Rapid colony transformation

Equipment and reagents

- Fresh plate of E. coli colonies (Protocol 1)
- TFB transformation buffer (or CCMB or FSB; Tables 4–6)
- 12 ml snap-cap polypropylene tubes (Falcon 2059)
- SOC medium (Table 1)

Method

1. Pick several colonies (or a clump of cells) off a plate using a tungsten inoculating loop or a wooden applicator stick, being careful to take no agar along with the cells.

2. Disperse the colonies in 200 μl of chilled standard transformation buffer (TFB) by vigorous vortexing or by repeated pipetting (CCMB or FSB can be used instead of TFB).

3. Incubate the cells on ice for 10 min.

4. Add the DNA solution (10–1000 ng) in <10 μl, swirl to mix, and incubate on ice for 10 min.

5. Heat shock the cells at 37–42°C for 90 sec. (This step is optional if > 100 ng of DNA is used.)

6. Add 400–800 μl SOC medium, and incubate 20–60 min at 37°C. (This step is also optional and unnecessary if > 100 ng of DNA is used in the transformation.)

Protocol 8. *Continued*

7. Plate several fractions (e.g. 1%, 5%, 25%) on SOB agar plates containing appropriate selective medium (*Protocol 3*). Incubate at 37 °C to establish colonies.

5. Induction of competence with electroshock treatment

The treatment of cells with a brief pulse of high voltage electricity has been found to permeabilize them towards entry of a variety of macromolecules (9, 10). It is presumed that the discharge of a voltage potential across an electric field in which cells are suspended transiently depolarizes their membranes and induces pores that can serve as entry points for macromolecules. Such electroshock treatment has been recently shown to be applicable to competence induction of *E. coli* and other bacteria. For *E. coli*, electroshock transformation is the most efficient method available, and approaches the theoretical maximum of 100% cell transformation frequencies. Efficiencies of greater than 10^{10} transformants/µg of pUC18 have been reported (11, 12), with more than 80% of all cells becoming transformed. More recent improvements in electronics, *E. coli* strains, and procedures have increased the transformation efficiency up to 5×10^{10} (13), as is described below.

The exposure of a dense suspension of bacterial cells and plasmid DNA to a high strength electric field of short duration induces DNA uptake and elicits competence for stable plasmid transformation. The electroshock is usually generated by the discharge of a high voltage capacitor through a mixture of bacterial cells and DNA suspended between two electrodes. The pulse length of the capacitor discharge can be varied by increasing the capacitor size, or the resistance in the circuit itself, which includes the mixture of cell suspension and DNA. A parallel resistor can also be used to modulate the resistance of the electroshock circuit. The time of the electric current pulse (the shock) is described by a decay time constant τ (Tau), which corresponds to the time at which the voltage has dropped to ~37% of its original value. The time constant of the electroshock is determined by the product ($\tau = R \times C$) of the resistance (both of the cell/DNA mixture and any parallel resistor) and the capacitance of the circuit through which the electric field is being discharged.

Field strengths used for optimal electroshock transformation of *E. coli* range from 12.5 kV/cm to 16.7 kV/cm (11, 13). The field strength is the maximum voltage applied to the cell suspension divided by the distance between the electrodes in which the cell suspension rests. The optimal combination of capacitor and resistor is different for each of these field strengths, but the decay time τ should be approximately five milliseconds for each.

Several other parameters other than the electroshock itself effect electroporation efficiencies. The *E. coli* strain MC1061 (7) and its derivative DH10B (8) give the highest efficiencies of electroporation (*Table 3*). Growth of cells in medium without added Mg^{2+} produces the highest competence. In addition, cells should be washed extensively to remove all salts, and the final cell slurry should be at a density of 5×10^{10}–1×10^{11} cells/ml, with an $OD_{550} >$ 200 units.

Protocol 9. Electroshock transformation of *E. coli*

Equipment and reagents

- Fresh *E. coli* colonies on a plate of **SOB−Mg agar**—(SOB−Mg medium (*Table 1*) plus 1.5% agar is autoclaved and poured at 50 °C into plastic Petri dishes)
- SOB−Mg medium (*Table 1*)
- Dedicated 2.8 litre Fernbach non-baffled shake flask, capped with a cotton plug; cleaned and sterilized (see Section 2.4)
- Dedicated 500 ml polypropylene centrifuge bottles with caps and seals (e.g. Curtin

- Matheson Scientific Cat. No. 249−165 for a Sorvall GS3 rotor), cleaned and sterilized as for the shake flasks (see Section 2.4)
- EWB buffer, at 4 °C: EWB is 10% re-distilled glycerol (Gibco-BRL) in ultrapure water, sterile filtered
- SOC recovery medium (*Table 1*)
- SOB agar plates containing appropriate antibiotic

A. *Preparation of frozen electrocompetent cells*

1. Pick a single 2–3 mm colony from a freshly streaked SOB−Mg agar plate and disperse it in 1 ml of SOB−Mg by vortexing. The cells are best streaked from a frozen stock or a fresh stab about 16–20 h prior to initiating liquid growth.

2. Inoculate the cells into a 500 ml Erlenmeyer flask containing 50 ml of SOB−Mg.

3. Incubate 37 °C at 275 r.p.m. overnight (preferably < 12 h).

4. The following day use 7.5 ml of this fresh overnight culture to inoculate 750 ml SOB−Mg in the Fernbach flask.

5. Incubate at 37 °C with moderate agitation until an OD_{550} of 0.75–1.0 units is reached (3–6 × 10^8 cells/ml).

6. Collect the cell suspension into chilled 500 ml centrifuge bottles.

7. Pellet the cells by centrifugation at 2700 *g* in a standard floor centrifuge for 12–15 min at 4 °C. Carefully decant the supernatant and discard.

8. Resuspend the cell pellet in an equal volume (to the original) of chilled 4 °C EWB. Dispersing the pelleted cells requires vigorous agitation by vortexing, or sharp rapping against a solid table. Maintain the cells on ice at this step and in steps **9–13**.

9. Centrifuge again to pellet the cells, and immediately decant the supernatant. (Cell loss is difficult to avoid but should be minimized.)

10. Resuspend the cell pellet as in step **8** and again centrifuge to pellet the cells. Thoroughly decant the supernatant after the centrifugation.

Protocol 9. *Continued*

11. Resuspend the cells with the few drops of excess liquid remaining in the tube by scraping the pellet with a sterile 1 ml pipette.

12. Measure the volume of the cell suspension. Then dilute a small portion (e.g. 1%) of the cell suspension 300-fold into SOB−Mg medium, and determine its relative OD_{550} using a spectrophotometer.

13. Calculate the absorbance of the concentrated cell slurry (multiplying the diluted value by 300), and adjust its volume as necessary with cold EWB to produce a final OD_{550} of 200–250 units/ml. If the undiluted OD_{550} is below these values, centrifuge the cells again, and repeat steps **11–13**.

14. Dispense the cell suspension in 120 μl aliquots into cold 2 ml cryo-tubes, freeze by **partial submersion** in a freshly prepared dry ice/ethanol bath, and store at −80 °C.

B. *Electroshock transformation*

1. Thaw electrocompetent cells on ice, and aliquot 25–45 μl into chilled polypropylene microcentrifuge tubes, on ice.

2. Add 1 μl of DNA solution in 0.5 × TE.

3. Transfer individual aliquots of cell and DNA into chilled electroporation cuvettes. For GIBCO-BRL 0.15 cm cuvettes, suspend 20 μl between the electrodes; for 0.2 cm Bio-Rad cuvettes, use 40 μl.

4. Electroshock under optimal conditions for the strain. For DH10B, these values are 16.7 kV/cm, a resistance of 4000 ohms, and a capacitance of 2 μF. (An alternative, if 16.7 kV/cm cannot be achieved, is 12.5 kV/cm, 25 μF, and 200 ohms.)

5. Remove electroshocked cells immediately from the chamber with a micropipette, and place in a chilled Falcon 2059 tube on ice.

6. Add 1 ml of SOC recovery medium (20 °C). Incubate for 1 h at 37 °C, 225 r.p.m.

7. Plate varying fractions on agar containing medium that selects for the transformed cells (*Protocol 3*). Incubate at 37 °C to establish colonies. If the concentration of transformed cells is uncertain, plate small fractions (0.01%–1%), saving the remainder for subsequent plating to achieve the desired colony density. See *Protocol 4* for short and long-term storage conditions.

5.1 Purification of DNA for electroshock transformation

DNA used with electrocompetent cells should be free of phenol, ethanol, and detergents, as with the high efficiency chemically competent cells. In addition, it is important that the DNA solution being used for electroshock

1: Techniques for transformation of E. coli

transformation has a very low ionic strength, and thus a high resistance. The following protocols achieve this purpose.

Protocol 10. Precipitation purification of DNA for electroshock transformation

Reagents

- Yeast tRNA solution: 1 μg/μl tRNA (GIBCO-BRL) in sterile water
- 5 M ammonium acetate, pH 7.5, sterile filtered
- 0.5 × TE: 5 mM Tris–HCl pH 7.4, 0.1 mM EDTA

Method

1. For a typical 20 μl ligation reaction, add 5–10 μg tRNA, and adjust the solution to 0.3 M in ammonium acetate using a 5 M stock solution. The volumes may be scaled up according to the amount of DNA being prepared.

2. Add 2 vol. of ethanol.

3. Centrifuge at 4 °C at >12 000 g for 15 min. Remove the supernatant with a micropipette.

4. Add 60 μl of 70% ethanol (made with 30% ultrapure water, no added salts or buffers) and centrifuge for 15 min. Remove supernatant with a micropipette.

5. Allow the pellet to dry at room temperature.

6. Resuspend DNA in 0.5 × TE to a concentration of 10 ng/μl DNA. Use 1 μl per discrete transformation of 20 μl of cell suspension.

Alternatively, ligations and other DNA preparations can be microdialysed against 0.5 × TE as drops on hydrophobic membrane filters to remove salts and thereby reduce conductivity (14), as detailed in *Protocol 11*.

Protocol 11. Microdialysis purification of DNA for electroshock transformation

Reagents

- Millipore type VS filter, 0.0025 μm pore size, 1 cm diameter
- 0.5 × TE: 5 mM Tris–HCl pH 7.4, 0.1 mM EDTA

Method

1. Remove 20–50 μl of the DNA solution from its container and place it as a drop on top of a Millipore filter, type VS 0.0025 μm, that is floating on a pool of 0.5 × TE (or 10% glycerol), for example in a small plastic Petri dish.

Protocol 11. *Continued*

2. Incubate at room temperature for several hours.

3. Withdraw the DNA drop from the filter and place it in a polypropylene microcentrifuge tube. Use 1 μl in a discrete electroshock transformation.

The floating drop method is particularly advantageous for very small volumes (and low concentrations) of DNA, where alcohol precipitation could result in significant loss of material. The method assumes that the DNA concentration is already appropriate for direct use in the electroshock transformation procedure.

6. Strain recommendations for transformation of *E. coli*

6.1 General considerations

Three factors are important in choosing a strain for generating cDNA or genomic libraries in plasmid vectors. The first is the transformation efficiency of the particular strain, since the competence of the cells can affect the ability to isolate the clone of interest. In general, fewer transformation reactions with a ligation mix are required to generate a representative library if the cells are highly competent. The second consideration is possible destruction of the incoming plasmid DNA by the *E. coli* restriction systems (*hsdRMS* and *MDRS*). See ref. 3 for a discussion of the genetics of these systems. A third consideration is the stability of the DNA being cloned, as certain sequences are not stable in standard *E. coli* cloning hosts.

6.2 cDNA cloning in plasmid vectors

For most cDNA cloning applications, we recommend standard *recA* strains such as DH5α, HB101, or DH10B that can be rendered highly competent for transformation. However, some cDNA cloning procedures involve methylation of the DNA, which would motivate the use of a *MDRS⁻* strain in order to prevent possible restriction of the clone of interest by the methylation dependent *mcrA, mcrBC,* or *mrr* systems. For procedures involving methylation of the cDNA, *MDRS⁻ recA⁻* strains such as DH5αMCR and DH10B are recommended. If cDNA cloning is being performed using the mini-plasmid vector ΠVX (15, 16), then a strain such as MC1061/p3 or DH10B/p3 carrying the defective antibiotic resistance plasmid p3 should be used. Transformation of these strains with ΠVX restores ampicillin and tetracyclin resistance, providing the selection for transformed clones.

The genotypes and characteristic transformation efficiencies of several reliable and generally successful plasmid cloning strains are listed in *Table 7*. Strains being used for the generation of cDNA libraries in M13 or phagemid vectors are discussed in Section 6.5

Table 7. Recommended *E. coli* strains for generating cDNA and genomic libraries in plasmid vectors

Strain	Relevant genetic loci								Maximal competence (XFE: colonies/μg pBR322)		Reference
	recA[a]	hsdR	mcrA	mcrB	mrr	gyrA	deoR	lacZΔM15[b]	Chemical treatment[c]	Electroshock[d]	
HB101	−(13)	−	+	−	−	+	+	−	$0.5–2 \times 10^8$	$1–2 \times 10^{10}$	31
MM294	+	−	+	+	+	+	+	−	$1–3 \times 10^8$	$3–10 \times 10^9$	32
DH5	−	−	+	+	+	−	−	−	$2–10 \times 10^8$	$3–10 \times 10^9$	8
DH5α	−	−	+	+	+	−	−	+	$2–10 \times 10^8$	$3–10 \times 10^9$	8
DH5αMCR	−	Δ	−	Δ	Δ	−	−	+	$1–5 \times 10^8$	$1–5 \times 10^9$	8
MC1061	+	−	−	−	+	+	+	−	$1–5 \times 10^8$	$2–5 \times 10^{10}$	7
DH10B	−	Δ	−	Δ	Δ	+	−	+	$2–10 \times 10^8$	$1–3 \times 10^{10}$	8
STBL2	−	Δ	−	Δ	Δ	−	+	−	$1–10 \times 10^7$	$1–10 \times 10^8$	37

[a] Unless indicated all *recA*− alleles are *recA1*; Δ = complete deletion of noted gene.
[b] Complements *lacZ* α subunit carried in many vectors to produce active β gal protein.
[c] Range of values given for MC1061, DH10B and their derivatives are based on *Protocol 5*, whereas all other strains are rendered most competent with *Protocols 6* and *7*.
[d] These values are obtained using electrical field strengths of 16.6 kV/cm. (Some commercial electroporation devices cannot achieve this field strength, and therefore cannot produce these transformation efficiencies.)

6.3 Genomic cloning and plasmid rescue using plasmid vectors

Many of the same considerations used to select a host for constructing cDNA libraries also apply to the construction of genomic libraries. As with cDNA libraries, most but not all genomic sequences will be stable in a *recA⁻* host strain. However, in contrast to cDNA cloning, most eukaryotic genomic DNA as well as many prokaryotic genomic DNA sequences are methylated. Therefore the preferred strain for generating representative genomic libraries should have the *recA⁻* and *mcrA⁻* mutations, and contain a deletion eliminating *hsdR, mcrBC,* and *mrr* (17). The strains DH5αMCR and DH10B (*Table 7*) are recommended for genomic cloning.

6.4 Genetic and physiological stabilization of unstable clones

Several *E. coli* strains defective in one or more genes involved in DNA repair and recombination have been used to stabilize cloned sequences containing direct repeats or palindromes (reviewed in ref. 3). Most of these strains have not been well characterized and little data exists on their susceptibility to high efficiency transformation. Therefore we cannot recommend these strains as standard hosts for constructing cDNA or genomic libraries. However, such strains could be considered as supplements when constructing libraries provided that a standard cloning strain serves as the primary host. It is our current experience that no known *E. coli* strain is capable of stabilizing every recalcitrant DNA sequence inserted into a plasmid or lambda cloning vector. However, we and others have recently developed a derivative of *E. coli* JM109, called STBL2, which has proved unusually adept at stabilizing difficult DNA sequences, including a full-length copy of the Simian immunodeficiency virus proviral DNA, and a variety of plasmids containing direct repeats (37). The genetic basis of this capability is under investigation. The STBL2 strain, in conjunction with growth at 25–35°C (see below), is recommended for attempts to stabilize sequences that are rearranging in standard hosts.

There is an alternative or a complement to the use of mutant *E. coli* for stabilization of 'difficult' DNA sequences during molecular cloning and amplification. We have observed that many unfavourable sequences can be preferentially stabilized by growing transformed cells at lower temperatures (25–30 °C) in very rich medium, and by not allowing such cultures to reach saturation. Thus, if a DNA sequence being cloned is either not appearing at all in the transformed cells, or is rearranging during amplification, we recommend the following:

(a) Incubate the cells following transformation in SOC medium (*Table 1*) at 30 °C.

(b) Establish colonies at 30 °C on agar plates composed of SOB medium, and expand individual colonies in TB medium (*Table 2*) at 30 °C.

(c) Collect the cells for DNA isolation mid to late in the period of logarithmic growth.

6.5 Strain recommendations: M13/phagemid cloning

Several factors are important in selecting a strain for transformation with filamentous bacteriophage DNA. The first is the transformation efficiency of the strain. The second is the yield of single-stranded (ss) M13 or phagemid DNA. The third is the stability of the insert. Most of the strains listed in *Table 8* can be transformed with efficiencies of $> 10^8$ transformants/μg M13 RF DNA. Among the recommended strains for M13 cloning are the *recA* strains JM109, DH5α/F′, and DH5α/F′IQ. In addition, most of these strains yield comparable quantities of ss M13 DNA. However, the yields of ss phagemid DNA can vary depending on the strain, and the ss DNA can be contaminated with double-stranded plasmid and chromosomal DNA resulting from cell lysis. To circumvent the problem of ds DNA contamination we recommend the *recA*⁻ strains DH11S (18, 19), derived from NM522, and DH12S (20), derived from DH10B. Both DH11S and DH12S are *endA*⁺ and therefore produce its gene product, a double-strand specific DNase, which is released via the periplasmic space, thereby rendering preparations of ss phagemid DNA from the medium virtually free of contaminating ds DNA (19). DH12S is the recommended strain for electroshock transformation, as it is much more efficiently transformed by this procedure (20).

6.6 Sources of *E. coli* strains

There are two primary repositories of *E. coli* strains for distribution to the scientific community. The most extensive collection is that of the *E. coli* Genetic Stock Center (ECGS), which is located at Yale University (Department of Biology, 255 OML, PO Box 6666, New Haven, CT 06510–7444, USA). A more limited collection is distributed by the American Type Culture Collection (12301 Parklawn Drive, Rockville, MD 20852–1776, USA). The noted *MDRS* mutants can be obtained from the authors at Life Technologies, Inc., 8717 Grovemont Circle, PO Box 6009, Gaithersburg, MD, 20884–9980, USA. In addition it is notable that an increasing number of *E. coli* strains are commercially available as preparations of frozen chemically competent or electrocompetent cells.

6.7 Optimization of competence for new *E. coli* strains

If the ultimate goal of a transformation experiment requires the use of a strain of *E. coli* that has not been described here, nor in our previous reports (3, 4, 6), it may be necessary to determine which transformation protocol is most suited for the strain. We recommend that trial transformations first be performed on the new strain using *Protocol 5* (CCMB based) and *Protocol 7* (TFB based).

27

Table 8. Recommended *E. coli* strains for DNA cloning using M13/phagemid vectors

Strain	hsdR	recA	Genes carried by F' plasmid	Other markers	Maximal chemical competence (XFE: colonies/μg pBR322)	Reference
JM109	–	–	traD36 proAB$^+$ lacI^q lacZΔM15		1–5×10^8	33
DH5α/F'	–	–	(No markers on F')	lacZΔM15	1–5×10^8	34
DH5α/F'IQ	–	–	proAB$^+$ lacI^q lacZΔM15 Tn5 (Km^r)		1–5×10^8	35
CJ236	+	+	pCJ105 (Cm^r)	dut 1 ung1	0.1–1.0×10^8	36
DH11S	Δ	Δ	proAB$^+$ lacI^q lacZΔM15	endA$^+$, mcrA, Δ(mrr, hsdRMS, mcrB)	0.2–0.5×10^9	18, 19
DH12S	Δ	–	proAB$^+$ lacI^q lacZΔM15	endA$^+$, mcrA, Δ(mrr, hsdRMS, mcrB)	0.5–1.0×10^{10} (electrocompetent)	20

For the CCMB protocol, cells should be grown without added magnesium (in SOB−Mg), whereas for the TFB based version, complete SOB medium is recommended. In addition, with the TFB version, the procedure can be tried without DMSO, without DTT, or without both. If neither of these protocols gives adequate competence, other variations on transformation buffers and procedures are given in our previous reports on this methodology (3, 4, 6). Finally, electroshock transformation can be assessed, using *Protocol 9*, or using a more general optimization protocol presented in ref. 3. It is our experience that most *E. coli* strains can be rendered competent to at least 10^7 transformants/µg, given a serious optimization process. However, unless the intended purpose specifically requires a new strain, those recommended in *Table 7* should be considered, as they have proven to be reliable hosts for plasmid transformation.

7. Parameters and evaluation of plasmid transformation

There are several criteria which are relevant to assessing the success of a competence induction protocol, and its suitability for the intended transformation. The results of a transformation can be considered from the perspective of the plasmid, or of the cell. Namely, one can either ask what is the probability that a given plasmid molecule will effect a transformed cell, or ask what fraction of the viable cells are competent for transformation? The first parameter is often referred to as the transformation efficiency (XFE), and is given in transformed colonies formed per unit mass of DNA (e.g. transformants per microgram of pBR322). It is more properly considered as the molecular transformation frequency, expressed as the probability that a plasmid molecule of a given size will transform a cell. The second parameter, the fraction of competent cells (F_c), illustrates the fact that not every cell becomes competent for transformation, which means that a given aliquot of competent cells can only produce a limited number of transformed cells, no matter how much DNA is added. This latter parameter becomes important when plasmid libraries of maximal number are the intended outcome of a transformation experiment. In most cases the primary parameter is the transformation efficiency, since this measures the amount of DNA which must be transformed to produce the desired number of colonies.

Another parameter of note is the growth state and cell density of the culture that was collected for preparing the competent cells. Cell density significantly influences subsequent induction of competence, and therefore can provide feedback for troubleshooting preparations with suboptimal competency. Other factors which can affect a transformation include competition from non-transforming DNA molecules, and effects of size and form of the transforming DNA. These parameters are described briefly below.

29

7.1 Determination of transformation efficiency

The transformation efficiency should be measured under cell excess, such that there is no competition between plasmids for the same competent cell. The traditional standard is the plasmid pBR322, although smaller plasmids based on the pUC vector are also used. There are about 2×10^{11} molecules of pBR322 per microgram, and about 3×10^{11} molecules of pUC. The number of viable cells used in an individual transformation ranges from 2×10^8 to 1×10^{11}, and thus the cells will always be in excess ($> 10:1$) if the control transformation uses less than 2×10^7 DNA molecules, which corresponds to 100 pg of pBR322. Values for XFEs range from 10^4 for rapid colony transformations to over 2×10^{10} transformed colonies per microgram of pBR322 with electrocompetent cells. Thus, with electrocompetent cells, up to 10% of the plasmid molecules are producing transformed cells. The amount of plasmid DNA used as a standard should be adjusted for the expected range of XFEs.

(a) For high efficiency transformations, 10 pg is a reliable amount for a control transformation. Following the transformation, different fractions of the transformation should be plated on to drug selection plates: 10%, and 1% (and 0.1% for electrocompetent cells). The 1% plating of 10 pg pBR322 will have 10 colonies if the XFE is 10^8, and 200 colonies if the XFE is 2×10^9.

(b) For electrocompetent cells, again use 10 pg of plasmid DNA, and after the electroshock and recovery, dilute 100 µl of the 1 ml transformation into 900 µl SOC, aliquot 10 µl of this dilution into a puddle of medium on an agar plate, and spread. This 0.1% dilution of a 10 pg transformation will produce 10 colonies if the XFE is 10^9, 100 if the XFE is 10^{10}, etc.

(c) For less efficient competent cell preparations, and also for assessing linearity of the response in highly competent cells, 1 ng of pBR322 is a reasonable amount for a second standard. The 1% plating of 1 ng will give 10 colonies if the XFE is 10^6, and 100 if the XFE is 10^7. For more efficient cells, perform serial dilutions of 10 µl of transformed cells into 1 ml SOC, plating 1% and 10%, etc. To assess the rapid colony transformation procedure, use 10 ng, plating 10%, which will give about 10 colonies if the XFE is 10^4.

We recommended that one control plasmid transformation be included on every occasion that a series of transformations is performed. The XFE gives important information on the state of the competent cells in the experiment. For example, if the experimental DNAs give no colonies, but the control XFE is as expected, then the experimental DNA is suspect. On the other hand, if the XFE is low, then the failure of the experiment may be explainable by inadequate competence, with a clear remedy being the acquisition of more efficient competent cells. Without this control there is no guidance into the problem.

7.2 Fraction of competent cells

The fraction of viable cells (F_c) that are competent is not unity, and as a result this parameter can influence the colony forming potential of an individual transformation, which in turn can determine the number of discrete transformations that need be performed in order to generate the desired number of transformants. The fraction of competent cells is traditionally measured with saturating levels of DNA, for example at cell to plasmid levels of $1:10$, which is the inverse of that employed to assess the XFE.

(a) Use 500–1000 ng of pBR322 DNA to determine F_c. Following transformation, the cells should be serially diluted by aliquoting 10 μl into 1 ml of SOC, and 10 μl of that into another 1 ml to give a 10^4 dilution. For chemically competent cells, where 2–5×10^8 cells are used in a transformation, 10 and 100 μl of this dilution are plated on to SOB agar (without antibiotic) and drug selection plates (10^5 and 10^6 final dilutions, respectively). The ratio of colonies formed on selective versus non-selective plates gives F_c.

(b) For electrocompetent cells, use 10 μg of plasmid DNA, and then take an additional $100 \times$ dilution, such that the colony number per plate is easily counted.

The fraction of competent cells ranges from ~0.001% for the rapid colony protocol, to 5–10% with the high efficiency chemical protocols, and up to 80% with the electroshock method. Moreover, when analysed below saturation, the XFE shows that electrocompetent cells are more efficient at producing transformation of a cell by a single plasmid molecule. Thus both XFE and F_c are important for maximizing the number of transformants actually produced in an experimental transformation. When crucial transformations are envisioned, such as for cDNA or genomic DNA library construction, both parameters should be assessed. Because electrocompetent cells have a ten-fold higher XFE and a higher F_c, they can be expected to produce 10–$100 \times$ more colonies than the chemical methods from the same amount of DNA under non-saturating conditions.

7.3 Saturation and optimization of colony forming potential

For large scale transformations, it is often valuable to titre the experimental DNA to maximize the colony forming potential for each discrete transformation, while avoiding saturation of the cells with excessive DNA, which might produce double transformants and waste precious DNA. Thus, while F_c is secondary in importance to the XFE, it does influence the overall colony forming potential of a competent cell aliquot. To determine the transforming potential of a DNA preparation, add increasing amounts of that DNA in a series of transformations. The point at which the total number of transformed colonies produced begins to level off is close to the optimum both for the number of transformants produced per individual transformation, and for the most conservative use of the DNA. In practice the optimum often turns out to be in the range where the number of DNA molecules has just exceeded the number of cells. For many cDNA transformations using chemical induction of competence, this optimal range has proved to be between 5–20 ng of ligated plasmid DNA.

7.4 Competition from non-transforming DNA

In many cloning situations only a fraction of the DNA will comprise plasmids which are capable of effecting transformed cells. The non-transforming DNA competes with the plasmids for DNA uptake (6). However, with chemical transformation there appear to be on the order of 100 separate channels into a cell, and with electroshock induction of competence it seems likely that a similar or even greater number of plasmid uptake channels (pores) are formed. Thus non-transforming DNA competes as it is able to occupy a majority of such channels on every competent cell. One can approximate the competition with the molar excess of DNA molecules over the total number of cells (since all cells appear to compete for DNA). Typically this competition becomes evident at about 100 ng of DNA, and becomes pronounced at 500 ng of DNA per discrete chemical transformation of $\sim$2–5 $\times$ 10^8 cells. With electroshock transformation, competition is not likely to be a factor, since the concentration of total DNA used for the transformation (of which 1 μl is combined with 20 μl of cells) would need to be in excess of 5 μg/μl before competition becomes significant for 2 $\times$ 10^9 cells. The effects of different types of non-transforming DNA cannot be predicted unambiguously, but in general, restricting the total DNA mass per unit transformation to less than 500 ng when non-transforming DNA is in excess proves to be a good approximation (6).

7.5 Effects of plasmid size and form on transformation

The extensive characterization of the properties of chemical transformation of *E. coli* (6) has revealed that very large plasmids, up to 66 kb, can be efficiently transformed into strains such as DH5α, using the high efficiency chemical transformation method (*Protocols* 6 and 7). If one considers the molecular transformation probability, i.e. the probability that a plasmid molecule will affect a transformed cell, then transformation frequency declines linearly with increasing size. If one instead considers the XFE, which is scaled as mass and not molecules, then the transformation efficiency declines with the inverse of the mass squared. A similar relation exists for electroshock transformation (13). The discovery that the *deoR* mutation selectively improves transformation of large DNA molecules (D. Hanahan, unpublished results) will facilitate the transformation of plasmids used for genomic cloning of large DNA fragments. Strains such as DH5α, DH5αMCR, and DH10B carry the *deoR* mutation, and the latter two also lack the methylation dependent restriction activities.

A second relevant property of transformation is that there is no significant requirement for DNA supercoiling in plasmid transformation (6). The fact that relaxed DNAs transform at similar frequencies to supercoiled forms is significant for cloning experiments involving the ligation of large DNAs.

While closed circular double-stranded plasmids transform *E. coli* readily, neither linear plasmids nor single-stranded DNA are effective at transformation. It is clear from work on integrative transformation that *recBC* (21, 22) or *recD* (23) mutants are dramatically better than *rec*[+] or *recA* strains for linear DNAs. However, even in a *recBC* strain background, transformation by linear plasmids is significantly reduced relative to closed circular forms (21). This result may relate to the other exonucleases present as part of the recombination and repair system (see below). Similarly, single-stranded phagemid DNA transforms *E. coli* at frequencies about 10^4 lower than double-stranded forms using either electroshock or chemical transformation methods.

8. Discussion, applications, and further information

The techniques described above present the best current methods for introducing plasmids into *Escherichia coli*, and we have provided guidance into their relative benefits. It is clear that electroshock transformation is the most efficient method, producing not only the highest transformation efficiencies, but also the highest fraction of competent cells. However, this technique requires an electroporation device and special cuvettes, in addition to the electrocompetent cells. The advent of small electroporation devices dedicated to *E. coli* electroshock transformation improves the convenience, as does the availability of commercial preparations of electrocompetent cells. Nevertheless, chemically competent cells provide convenient and appropriate competent *E. coli* for a wide spectrum of DNA cloning procedures. *Table 6* summarizes the relative efficiencies and typical applications of the various procedures presented in this chapter. The generation of high complexity plasmid libraries is readily achievable with both the high efficiency chemical competence protocols and the electroshock transformation procedure.

A recent comparative study suggests that plasmid cloning with electroshock transformation of DH10B is as efficient, if not more efficient, than bacteriophage lambda based methods (24). The continuing popularity of λ vectors rests in part on the ease of manipulating large libraries of clones for amplification, screening, and storage/distribution. A reliable technique for manipulating and storing large plasmid libraries as primary distributions on membrane filters has been described previously (25–27). While this method preserves the primary complexity of a library, it is laborious. However, two simple techniques have been developed for easily amplifying, storing, and distributing plasmid libraries much as for bacteriophage libraries. These new methods serve to alleviate this common complaint about utilizing plasmids as vectors for high complexity libraries. One method disperses the transformed *E. coli* cells in a matrix of semi-soft agar and allows colonies to form, after which the amplified collection of transformed cells can be readily collected and aliquoted for freezing and/or plated for screening (28). This procedure is easy, and appears to normalize all of the colonies to approximately the same size, irrespective of differential growth rates of individual clones on an agar surface or in liquid culture. The second procedure involves amplification of colonies at high density on nylon or nitrocellulose filters laid on agar plates, followed by scraping the filters in SOB/glycerol medium to combine the collection of clones into a suspension that can be aliquoted and distributed, much as in a λ plate amplification. That protocol can be found in our more comprehensive review of plasmid transformation (3), which also addresses the transformation of other bacteria, and discusses genetic factors influencing plasmid transformation of *E. coli*. The reader is referred to this review for such information.

The cloning of large genomic DNA fragments into *E. coli* has generally been restricted to λ and cosmid vectors, which allow efficient DNA transfer and a selection for large insert size via *in vitro* packaging reactions to produce infectious virions containing recombinant phage or cosmids. This methodology, however, is limited to cloning fragments in the range of 20–40 kb by the phage λ headful packaging size constraint. Recently, larger genomic fragments (up to 100 kb) have been introduced into *E. coli* using phage P1 as a vector, the 'PAC' technology (29). Although direct plasmid transformation can also introduce such large plasmids, its application has been limited by the poor molecular transformation frequencies obtained with chemical competence techniques (6). With the advent of high efficiency electroshock transformation protocols, the possibility of plasmid based cloning of large DNA fragments has been re-assessed, and a new vector system developed (30). These vectors have been shown to stably carry 300 kb fragments; the recombinants are referred to as bacterial artificial chromosomes (BAC). The BAC vectors are based on the F factor origin of replication, and therefore are maintained at one copy per cell, thus acting more like a genuine chromosome, without the pressure of high copy number plasmid replication. Electroshock transformation can readily introduce BACs into *E. coli*, raising the possibility that *E. coli* could well serve as a host for large scale chromosomal DNA cloning and genomic physical mapping projects.

References

1. Mandel, M. and Higa, A. (1970). *J. Mol. Biol.*, **53**, 159.
2. Cohen, S., Chang, A., and Hsu, L. (1972). *Proc. Natl. Acad. Sci. USA*, **69**, 2110.
3. Hanahan, D., Jessee, J., and Bloom, F. (1991). In *Methods in enzymology* (ed. J. H. Miller), Vol. 204, pp. 63–113. Academic Press, New York.
4. Hanahan, D. (1985). In *DNA cloning techniques: a practical approach* (ed. D. Glover), pp. 109–35. IRL Press, London.
5. Tartof, R. and Hobbs, C. (1987). *BRL Focus*, **9(2)**, 12.
6. Hanahan, D. (1983). *J. Mol. Biol.*, **166**, 557.
7. Casadaban, M. and Cohen, S. (1980). *J. Mol. Biol.*, **138**, 179.
8. Grant, S. G. N., Jessee, J., Bloom, F. R., and Hanahan, D. (1990). *Proc. Natl. Acad. Sci. USA*, **87**, 4645.
9. Knight, D. (1981). *Tech. Cell. Physiol.*, **113**, 1.
10. Neumann, E., Schaefer-Ridder, M., Wang, Y., and Hofschneider, P. (1982). *EMBO J.*, **7**, 841.
11. Dower, W., Miller, W., and Ragsdale, C. (1988). *Nucleic Acids Res.*, **16**, 6127.
12. Calvin, W. and Hanawalt, P. (1988). *J. Bacteriol.*, **170**, 2796.
13. Smith, M., Jessee, J., Landers, T., and Jordan, J. (1990). *BRL Focus*, **12(2)**, 38.
14. Jacobs, M., Wendt, S., and Stahl, U. (1990). *Nucleic Acids Res.*, **18**, 1653.
15. Seed, B. (1983). *Nucleic Acids Res.*, **11**, 2427.
16. Aruffo, A. and Seed, B. (1987). *Proc. Natl. Acad. Sci. USA*, **84**, 8573.
17. Woodcock, D. M., Crowther, P. J., Doherty, J., Jefferson, S., DeCruz, E., Noyer-Weidner, M., Smith, S. S., Michael, M. Z., and Graham, M. W. (1989). *Nucleic Acids Res.*, **17**, 3469.
18. Lin, J. J., Smith, M., Jessee, J., and Bloom, F. (1991). *BRL Focus*, **13(3)**, 96.
19. Lin, J. J., Smith, M., Jessee, J., and Bloom, F. (1992). *BioTechniques*, **12(5)**, 718.
20. Lin, J. J., Jessee, J., and Bloom, F. (1992). *BRL Focus*, **14(3)**, 98.
21. Conley, E. and Saunders, J. (1984). *Mol. Gen. Genet.*, **194**, 211.
22. Winans, S., Elledge, S., Krueger, J., and Walker, G. (1985). *J. Bacteriol.*, **161**, 1219.
23. Shevell, D., Abou-Zamzam, A., Demple, B., and Walker, G. (1988). *J. Bacteriol.*, **170**, 3294.
24. Gruber, C. E. (1992). *BioTechniques*, **12(6)**, 805.
25. Hanahan, D. and Meselson, M. (1980). *Gene*, **11**, 63.
26. Hanahan, D. and Meselson, M. (1983). In *Methods in enzymology* (ed. R. Wu, L. Grossman, and K. Moldave), Vol. 100, pp. 333–42. Academic Press, New York.
27. Hanahan, D. and Meselson, M. (1989). In *Recombinant DNA methodology; selected methods in enzymology* (ed. R. Wu, L. Grossman, and K. Moldave), pp. 267–76. Academic Press, New York.
28. Kriegler, M. (1990). In *Gene transfer and expression: a laboratory manual*, pp. 131–3. Stockton Press, New York.
29. Sternberg, N. (1990). *Proc. Natl. Acad. Sci. USA*, **87**, 103.
30. Shizuyu, H., Birren, B., Kim, U.-J., Mancino, V., Slepak, T., Tachiri, Y., and Simon, M. (1992). *Proc. Natl. Acad. Sci. USA*, **89**, 8794.
31. Boyer, H. W. and Roulland-Dussoix, D. (1969). *J. Mol. Biol.*, **41**, 459.
32. Meselson, M. and Yuan, R. (1968). *Nature*, **217**, 1110.

D. Hanahan, J. Jessee, and F. R. Bloom

33. Yanisch-Perron, C., Vieira, J., and Messing, J. (1985). *Gene*, **33**, 103.
34. Liss, L. R. (1987). *BRL Focus*, **9(3)**, 13.
35. Jessee, J. and Blodgett, K. (1988). *BRL Focus*, **10(4)**, 69.
36. Kunkel, T. A., Roberts, J. D., and Zakour, R. A. (1987). In *Methods in enzymology* (ed. R. Wu and L. Grossman), Vol. 154, pp. 367–82. Academic Press, New York.
37. Trinh, T., Jessee, J., Bloom, F. R., and Hirsch, V. (1994). *BRL Focus*, **16**, 78.

2

Construction of representative genomic DNA libraries using phage lambda replacement vectors

KIM KAISER, NOREEN E. MURRAY, and
PAUL A. WHITTAKER

In the years since this article first appeared, the technology it describes has become much more routine. This is in large part due to the contribution that the biotechnology industry has made to the provision of reliable reagents, ready-prepared vector DNA (cleaved and dephosphorylated if necessary), packaging extracts, host strains, and even kits to take one through the entire procedure from beginning to end. Failing all else one can send away a tissue sample and receive, if at some cost, a library in the post. Not everybody works in such a privileged environment, however, and not everybody wants the standard product. It thus remains useful to know how to generate a library from first principles. Basic protocols are given in this article. It also remains important to understand the biology behind library construction and propagation, as the knowledge may help one to overcome problems posed by the occasional anomalous result. Such matters are also dealt with. For more general information, we would advise the reader to consult refs 1, 2 (which has a useful lambda methods section), and 3.

1. The perfect library

A perfect genomic DNA library would contain DNA sequences representative of an entire genome, in a stable form, as a manageable number of overlapping clones. The cloned fragments would be sufficiently large to contain whole genes and their flanking sequences, but not too large to hinder mapping by restriction enzyme cleavage. Libraries should be both easy to construct from small amounts of starting material, and easy to screen for the sequences of interest, usually by hybridization with labelled DNA or RNA probes. Overlapping clones facilitate 'walking' into adjacent regions of the genome, and simple means should be available for generating vector-free probes representing the ends of inserts. Finally, it should be possible to amplify a library without loss or misrepresentation, and to store it for years without significant decrease in titre.

Several different cloning systems are currently used for the construction of genomic DNA libraries. These range from bacteriophage lambda (λ) vectors, with a maximum insert length of little more than 20 kb, to yeast artificial chromosome (YAC) vectors with up to 1 Mb capacity (4). Cosmid (1) and bacteriophage P1 (5) vectors cover the middle range (≈ 40 kb and 100 kb respectively). Though each system has its merits, lambda vectors still provide the simplest and most efficient means of constructing libraries of general utility.

2. Basic principles

We introduce the basic technicalities of library construction with minimal reference to phage biology. Later sections discuss aspects of the latter that are relevant to a more informed use.

The wild-type lambda genome is a 48.5 kb linear DNA molecule (6), only 60% of which encodes essential functions. It is equipped with complementary single-stranded termini, cohesive ends, which allow formation of circular and multimeric (concatemeric) derivatives. As the result of many *in vivo* and *in vitro* manipulations, lambda has given rise to an astonishing variety of vectors for the propagation of foreign DNA (1–3, 36). Here we are concerned only with the class known as replacement vectors (see Section 14).

The basic strategy for use of a replacement vector is illustrated in *Figure 1*. Cleavage of vector DNA with an appropriate restriction enzyme generates three fragments: a left arm and a right arm together containing all the information for production of an infective phage particle, and a dispensable 'stuffer' fragment. Ligation in the presence of cleaved donor DNA generates novel DNA molecules, a proportion of which contain a donor DNA fragment flanked by left and right arms. The stuffer has been **replaced**. Introduced into *E. coli*, such novel molecules can enter the lytic growth cycle and produce a clone of viral particles that is recognized as a plaque (a zone of lysed bacteria). Only genomes in the 40–52 kb range are packaged into phage particles with high efficiency, and two arms alone generate a DNA molecule much shorter than 40 kb. Ligation in the absence of the stuffer thus provides a powerful enrichment for recombinant genomes.

Although plaques can be obtained following introduction of phage DNA directly into *E. coli* (transfection), only ≈ 10^4 recombinant clones per microgram of donor DNA are likely to be obtained. This might just suffice for a library representing a small genome. Considerably higher efficiencies are preferable for larger genomes, both for reasons of scale and so that plaques can be obtained at high density. Greater than 10^7 recombinants per microgram of donor DNA can be generated by *in vitro* packaging of recombinant DNA molecules before introduction to the bacteria. The preferred substrate

10 kb

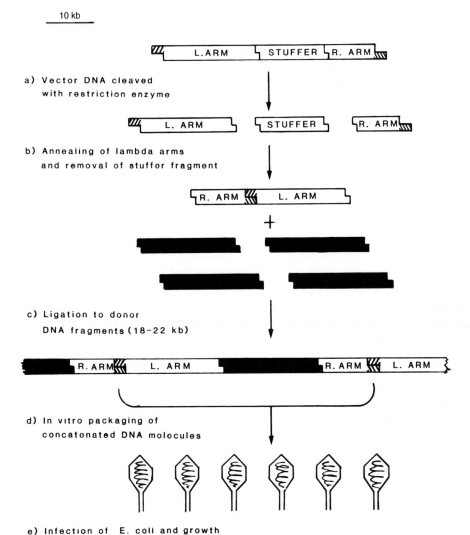

a) Vector DNA cleaved
   with restriction enzyme

b) Annealing of lambda arms
   and removal of stuffer fragment

c) Ligation to donor
   DNA fragments (18–22 kb)

d) In vitro packaging of
   concatenated DNA molecules

e) Infection of E. coli and growth
   of recombinant phage as plaques

Figure 1. Basic strategy for cloning using a lambda replacement vector. Cleavage of vector DNA with the appropriate enzyme allows it to accept donor DNA fragments with complementary termini. Hatched boxes denote the lambda cohesive ends. As shown here, they were allowed to anneal before ligation. The concatemeric ligation product is packaged *in vitro* to give infective phage particles, each of which is recovered in an amplified form as a plaque on the surface of an agar plate. A replica of the plaque distribution can be obtained by 'blotting' the surface with a nitrocellulose or nylon membrane (Section 8), to which a probe DNA may then be hybridized.

for *in vitro* packaging is a concatemer in which a recombinant DNA molecule is flanked on either side by a *cos* site (hatched boxes in *Figure 1*), a sequence that includes the double-stranded version of a cohesive end.

High molecular weight donor DNA is generally reduced to a size appropriate for insertion into the vector by partial cleavage with a restriction enzyme. This generates a set of overlapping fragments more or less representative of the genome. Multiple insertion can be avoided by using a size fraction of donor DNA close to the maximum acceptable to the vector. Alternative strategies include dephosphorylation of donor DNA, and partial fill-in of the donor and vector termini generated by restriction enzyme cleavage (Section 3.2).

Unless the stuffer is removed, or otherwise discouraged from re-ligating to vector arms, the library will have a background of reconstituted vector. This can be minimized by physical separation of the stuffer and the arms, by dephosphorylation of cleaved vector DNA, or by other enzymatic means (Section 3.5). Many vectors allow a genetic selection against non-recombinant phage during propagation of a library (Section 13).

The efficiency with which recombinant genomes are recovered as plaques is influenced considerably and selectively by host genetic background. The generation of *E. coli* strains with minimal abilities to restrict, recombine, and rearrange foreign DNA has been one of the most significant developments of recent years (Section 15).

3. Donor and vector DNAs

3.1 Isolation of donor DNA

The single most important characteristic of the donor DNA, whatever its source, is that it be of high molecular weight. Some degree of breakage is inevitable due to mechanical shearing during isolation, and due to the action of nucleases released during rupture of cells. Fragments with one ragged end can ligate with only a single vector arm. They also inhibit formation of long chains of recombinant genomes, the preferred substrates for *in vitro* packaging (Section 4.1). As a rule of thumb the average size of donor DNA before cleavage must be ≥ 100 kb.

Most methods suitable for the isolation of high molecular weight DNA involve proteinase treatment of cells or nuclei that have been lysed with a strong detergent (e.g. SDS) in the presence of EDTA (1, 3). EDTA serves to inhibit endogenous DNase activity. Prior to lysis, mechanical methods such as homogenization are permissible. During lysis, viscosity rises due to the release of high molecular weight DNA. From this stage onwards it is essential that mechanical shearing (mixing, shaking, extrusion through pipettes, or syringe needles), and the number of steps likely to cause it, be reduced to a minimum. Deproteinization of a lysate by extraction with phenol inevitably involves some degree of agitation. Care should be taken to ensure that it is performed gently.

Protocol 1 provides DNA in the 100–150 kb size range from tissue culture cells. For white blood cells, fresh or frozen tissue, and sperm, cells must first be isolated and washed. Suitable protocols may be found in ref. 57.

Protocol 1. Isolation of high molecular weight DNA from tissue culture cells

Equipment and reagents

- Phosphate-buffered saline (PBS): 8 g NaCl, 0.2 g KCl, 1.44 g Na_2HPO_4, 0.24 g KH_2PO_4 in 800 ml distilled H_2O, adjust pH to 7.4 with HCl, and add distilled water to 1 litre. Dispense and sterilize by autoclaving for 20 min at 15 lb/in^2. Store at room temperature
- TE: 10 mM Tris–HCl pH 7.5, 1 mM EDTA
- Extraction buffer: 10 mM Tris–HCl pH 8.0, 0.1 M EDTA, 0.5% SDS
- Proteinase K
- Phenol containing 0.1% hydroxyquinoline: equilibrate with 0.1 M Tris–HCl pH 8.0/0.2% β-mercaptoethanol, until the pH of the aqueous phase is greater than 7.6

- Chloroform/isoamyl alcohol (24:1)
- 5 M ammonium acetate
- Isopropanol
- 70% ethanol
- Sterile 100 ml conical flasks
- Sterile screw-capped 50 ml polypropylene centrifuge tubes (Falcon 50 ml centrifuge tubes or equivalent)
- MSE Mistral 6L centrifuge (or equivalent)
- Water-bath or incubator at 50 °C
- 1.5 ml microcentrifuge tubes
- Spectrophotometer

Method

1. Pellet 10^8 cells at 1500 *g* for 10 min.

2. Resuspend in 5–10 vol. ice-cold PBS. Repeat centrifugation.

3. Repeat step **2**.

4. Resuspend thoroughly in 2 ml TE and pour slowly down the side of a 100 ml conical flask into 20 ml of extraction buffer. Swirl while pouring to minimize clumping.

5. Add proteinase K to a final concentration of 100 μg/ml. Incubate for 3 h at 50 °C. Swirl periodically.

6. Cool to room temperature and add an equal volume of phenol. Mix by swirling until an emulsion forms. Add an equal volume of chloroform/isoamyl alcohol and swirl to mix.

7. Gently pour emulsion into a 50 ml centrifuge tube. Separate phases at 2500 r.p.m. for 15 min at room temperature in the 6L centrifuge.

8. Gently transfer aqueous DNA phase (top) to a clean 100 ml conical flask using a wide bore pipette. To avoid disturbing material at the interface, draw the DNA slowly into the pipette.

9. Repeat steps **6** to **8** until interface is clear (up to five extractions).

Protocol 1. *Continued*

10. Transfer aqueous phase to a clean centrifuge tube and add an equal volume of 5 M ammonium acetate. Mix thoroughly, add an equal volume of isopropanol, and mix by inversion. The DNA should form a fibrous precipitate. Pick it up using a Pasteur pipette with a sealed and U-shaped end, and transfer it to a 1.5 ml microcentrifuge tube. If the precipitate is fragmented, recover by pelleting at 2500 r.p.m. for 5 min at room temperature in the 6L centrifuge.

11. Remove remaining isopropanol by washing twice with 1 ml of 70% ethanol. Dry the DNA briefly under vacuum, and resuspend in 1 ml TE. To aid resuspension, place on a rotary mixer at room temperature overnight.

12. Measure the absorbance of the DNA at 260 nm and 280 nm. The 260/280 ratio should be greater than 1.8 for DNA that has been effectively deproteinized. The DNA concentration can be determined from the absorbance reading at 260 nm (a solution with an OD_{260} of 1 contains approximately 50 µg DNA/ml). 300–400 µg of DNA should be recovered per 10^8 diploid human cells.

13. Analyse the DNA by electrophoresis at 2 V/cm in a low percentage ($\leq 0.5\%$) agarose gel with appropriate size markers (e.g. intact lambda DNA, 50 kb; intact T4 DNA, 165 kb). Alternatively, and more accurately, DNA can be sized using pulsed field gel electrophoresis.

Points to note in the preparation of donor DNA are:

(a) DNA of the highest molecular weight will be obtained from *fresh* cells or tissue. If it is not possible to process samples immediately, freeze them quickly by immersion in liquid nitrogen, and store them frozen at $-70\,°C$. Cut large pieces of tissue into small pieces with a pair of surgical scissors before immersion. This will assist both their rapid freezing and their eventual pulverization.

(b) Preparation of DNA from nuclei rather than from whole cells will minimize the contribution of organellar DNAs to a library.

(c) Very viscous DNA is generally difficult to deproteinize. It is better to work at a concentration of < 1 mg DNA per 20 ml of nuclear lysate, and to choose the amount of starting material accordingly. Pronase (Boehringer-Mannheim molecular biology grade) is a cheap alternative to proteinase K, and works just as well.

(d) Phenol extraction can be performed gently by swirling the mixture in a conical flask, and then pouring the emulsion into a centrifuge tube for separation of the phenol and DNA phases. Use of silica gel polymer simplifies separation (7). The gel forms an impenetrable barrier between the organic and aqueous phases. DNA can thus be poured, rather than aspirated, from the centrifuge tube.

(e) Isopropanol precipitation of extracted DNA is preferable to ethanol precipitation because RNA remains in solution. This usually eliminates the need for subsequent RNase treatment. Make the DNA 2.5 M in ammonium acetate, and then add an equal volume of isopropanol. Recover the precipitate by low speed centrifugation. To aid resuspension of precipitated DNA, mix gently on a rotary mixer for up to 24 hours.

(f) Electrophoresis at low voltages improves the resolution of high molecular weight DNA species (8).

(g) Sterile DNA solutions containing no active enzymes can be stored for long periods at 4 °C. High molecular weight genomic DNA should never be frozen.

3.2 Size fractionation

A truly random library requires fractionation of DNA in a sequence-independent manner, such as controlled mechanical shearing (9). It is usually more convenient to fractionate DNA by partial cleavage with a restriction enzyme. If the enzyme cuts frequently compared with the desired insert size, a set of overlapping fragments will be generated that is sufficiently random for most purposes. It cannot be guaranteed, however, that all regions of the genome are equally represented in the desired size range, since regions with an abnormal distribution of cleavage sites may occur, and every cleavage site may not be an equally good target.

Because of their compatibility with *Bam*HI (Section 3.3), the enzymes *Sau*3A and *Mbo*I are commonly used. Both recognize the sequence GATC, which occurs on average every 4^4 (256) bp in random sequence DNA. To generate 20 kb fragments it is necessary to cleave only $\approx 1/80$ of the available sites. Since the actual ratio will vary with base composition, it is usual to determine the conditions empirically in a series of small scale reactions (*Protocol 2*), and then to digest 150–300 μg of genomic DNA by scaling up the pilot conditions exactly. For the scale up, it is recommended that 50–100 μg samples of DNA are digested using three different concentrations of restriction enzyme that straddle the optimal concentration determined in the pilot experiment. At the end of the digestion, analyse 1 μg samples of the digested DNA from each of the three reactions by gel electrophoresis to ensure that all has gone according to the predictions of the pilot experiment (store the remainder on ice). If all has gone well, the three reactions can be pooled, gently extracted with phenol, ethanol precipitated, and resuspended in TE buffer (*Protocol 1*).

Note: cleavage with *Sau*3A can be blocked by cytosine methylation, cleavage with *Mbo*I by adenine methylation (10). Choice of enzyme may thus reflect the methylation status of ones donor DNA.

Protocol 2. Establishing conditions for partial digestion of
donor DNA

Equipment and reagents

- High molecular weight genomic DNA prepared as described in *Protocol 1*
- Sterile distilled water
- 1.5 ml microcentrifuge tubes
- Restriction enzyme and 10 × digest buffer
- 0.1 M EDTA pH 7.5
- Gel loading buffer: 25% Ficoll type 400,
0.25% orange G dye in water

- 0.5% agarose gel
- Ethidium bromide
- Camera with Wratten 22A filter
- UV transilluminator
- λ/*Hind*III and λ/*Sal*I size markers

Method

1. Prepare a reaction mixture containing 10 μg of DNA and 1 × restriction
   enzyme buffer in a final volume of 150 μl.

2. Label nine microcentrifuge tubes. Dispense 30 μl into tube 1 and
   15 μl into tubes 2–9.

3. Add 5 U of restriction enzyme to tube 1, mix thoroughly, and transfer
   15 μl to tube 2. Mix well and continue the twofold serial dilution
   through to tube 8 (tube 9 is the undigested control). The concentration
   of enzyme will vary from 2.5 U/μg DNA in tube 1 to 0.02 U/μg in tube 8.

4. Incubate tubes at 37 °C for 60 min.

5. Stop the reactions by adding 3 μl 0.1 M EDTA pH 7.5 per 15 μl reaction
   volume and 2 μl gel loading buffer.

6. Analyse by electrophoresis through a 0.5% agarose gel at 1–2 V/cm
   (*Figure 2*). Appropriate size markers are lambda DNA cleaved with
   *Hind*III and/or lambda DNA cleaved with *Sal*I.

7. Stain the gel with ethidium bromide (0.5 μg/ml for 30 min) and photo-
   graph without over-exposing the film.

8. Determine the amount of enzyme needed to produce the maximum
   intensity of fluorescence in the 15–25 kb size range. The intensity of
   fluorescence is related to the mass distribution of the DNA. To obtain
   the maximum number of molecules in a particular size range, use half the
   amount of enzyme that produces the maximum amount of fluorescence.

The size distribution after cleavage should peak in the 18–22 kb range.
There will inevitably be a proportion of smaller fragments, however, that if
not removed can become incorporated as multiple inserts. At the other
extreme, large fragments will give rise to genomes too long to be packaged.
Moreover, recombinant DNA molecules just at the upper and lower limits of
the packageable size range may be unduly susceptible to Rec-dependent
deletion and duplication, or Rec-independent deletion and rearrangement

λ   1   2   3   4   λH3

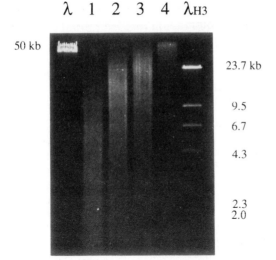

50 kb

23.7 kb

9.5
6.7

4.3

2.3
2.0

Figure 2. Partial cleavage of high molecular weight DNA with *Sau*3A. DNA samples cleaved at concentrations of enzyme embracing the optimum were separated by electrophoresis at 2 V/cm in a 0.3% agarose gel. The figure is a photograph of the gel after staining with ethidium bromide. The markers are intact lambda DNA and lambda DNA cleaved to completion with *Hind*III. They were heated to 68 °C for 10 min prior to electrophoresis in order to dissociate the cohesive ends. 200 µg of donor DNA was cleaved in subsequent large scale reactions, half under the conditions of lane 2 and half under the conditions of lane 3. Note that a portion of the DNA appears refractory to cleavage by *Sau*3A, and thus could not be cloned.

(Section 15.2). Fragments in the desirable range can be separated physically by centrifugation through sucrose or sodium chloride gradients (1, 3), or by electrophoresis in agarose gels (11, 12). The former procedures can provide DNA in good yield, but at the expense of a spread of fragment sizes within each fraction. They also require a lot of starting material. Gel purification gives tighter fractionation and requires less DNA, but many batches of agarose contain inhibitors of DNA ligase. Elutip columns (Schleicher and Schull) are very effective at removing ligase inhibitors from DNA.

A method for the size fractionation of partially digested genomic DNA using sucrose gradients is given in *Protocol 3*. Sodium chloride gradients allow reduced centrifugation times, but more care must be taken in preparing the gradients because of the lower viscosity of the solutions. Sucrose gradients can be prepared using a gradient maker, or by overnight diffusion at 4 °C of a step gradient. This is the method described in *Protocol 3*. If each sucrose 'step' is frozen before the next layer is added, the gradient can be stored frozen indefinitely. Frozen gradients can be used after thawing at 4 °C overnight. No more than 200 µg of genomic DNA should be loaded on a 38 ml sucrose gradient to prevent overloading. Centrifugation times can be reduced from 24 h to 3 h by using a vertical rotor (e.g. Beckman VTi 50) rather than a swing-out rotor.

Protocol 3. Sucrose gradient fractionation of partially digested donor DNA

Equipment and reagents

- Partially digested genomic DNA
- 10%, 15%, 20%, 25%, 30%, 35%, and 40% sucrose solutions made in a buffer containing 20 mM Tris pH 8.0, 100 mM NaCl, 10 mM EDTA (sterile filtered through a 0.2 μm nitrocellulose filter and stored at 4°C)
- Beckman L8 ultracentrifuge (or equivalent)
- 38 ml ultracentrifuge tubes (Beckman SW28 polyallomer or equivalent)
- Swing-out rotor (Beckman SW28 or equivalent)
- TE (*Protocol 1*)
- 3 M sodium acetate pH 6.0

- Dextran T500 (10 mg/ml made up in distilled water and autoclaved for 20 min at 15 lb/in^2)
- Ethanol
- 21 gauge syringe needles
- Peristaltic pump and tubing
- Microcentrifuge and 1.5 ml microcentrifuge tubes
- 0.5% agarose gel
- λ/*Hind*III and λ/*Sal*I size markers
- Gel loading buffer (*Protocol 2*)
- Camera with Wratten 22A filter
- UV transilluminator

A. *Preparation of sucrose gradients*

1. Pipette 5 ml of 40% sucrose solution into a 38 ml centrifuge tube.

2. Gently overlay with 5 ml of 35% sucrose solution.

3. Repeat step **2** with the remaining sucrose solutions.

4. Place at 4°C overnight to allow the sucrose steps to diffuse.

B. *Centrifugation and size fractionation*

1. Layer no more than 200 μg partially digested DNA in a volume of 500 μl onto each gradient.

2. Centrifuge at 26 000 r.p.m. for 24 h at 20°C in a Beckman SW28 rotor or equivalent.

3. Collect 1 ml fractions. Either insert a 21 gauge needle through the bottom of the centrifuge tube and drip, or pump/syphon the sucrose solution through a 100 μl glass capillary that has been carefully lowered to the bottom of the centrifuge tube.

4. Remove 18 μl from each of the first 20 fractions, add 2 μl of gel loading buffer, and analyse on a 0.5% agarose gel at 2 V/cm with lambda DNA size markers.

5. Photograph the gel and locate the fractions containing DNA fragments in the 15–25 kb range. To each fraction add dextran T500 to 200 μg/ml, sodium acetate pH 6.0 to 0.3 M, and 2 vol. of ethanol (do not pool the fractions at this stage). Mix well and place on ice for 10 min.

Collect the precipitated DNA by centrifugation in a microcentrifuge for 15 min at room temperature.

6. Remove ethanol and resuspend each pellet in 200 μl TE. Re-precipitate as in step **5**, but omit the dextran T500.

7. Resuspend each pellet in 20 μl TE. Analyse 1 μl on a 0.5% agarose gel (as step **4**) to check the size distribution.

8. Spot 1 μl of each gel fraction, together with 1 μl samples of vector DNA at known concentrations, on to a 1% agarose gel containing 1 μg/ml ethidium bromide (pour the gel in a Petri dish and allow the surface to dry). After the DNA has diffused into the agarose, photo-graph the gel and determine DNA concentrations by comparing inten-sity of fluorescence in the samples with that of the standards.

9. Fractions containing DNA fragments in the desirable size range can be pooled. Alternatively, determine the efficiency with which DNA in each fraction can be incorporated into packageable DNA molecules, and use this as a guide to which fractions to use for library construc-tion.

10. The size fractionated DNA can be stored frozen in TE for several years at −20 °C.

The range of molecular lengths in each fraction can be determined by electro-phoretic separation in a 0.3–0.5% agarose gel (*Figure 3*). Fractions that contain DNA of appropriate size are then concentrated by ethanol precipi-tation. Since indeterminate losses can occur during precipitation of small amounts of DNA, it is advisable to use a carrier. 200 μg/ml Dextran T500 (Pharmacia) is effective and has minimal effects on ligation or *in vitro* pack-aging. tRNA can inhibit the latter.

As an alternative to size fractionation, partially cleaved DNA may be treated with phosphatase and ligated directly to vector DNA (13). Since ligase is unable to join two DNA molecules if both are lacking a 5′ terminal phosphate group, treatment of donor DNA with phosphatase prevents multiple insertion. Unfortunately, it also precludes the use of dephosphorylated vector DNA, a preferred means of reducing non-recombinant background (Section 3.4). Another option is partial fill-in of donor and vector termini (14). For example, *Sau*3A/*Mbo*I ends partially filled with A and G will not ligate to themselves, but will ligate to *Xho*I cleaved vector DNA partially filled with C and T (Section 3.4). In either case there will be a proportion of recombinant DNA molecules too small or too large to be packaged, and it is left to the packaging extracts (Section 4.2) to discriminate against them. Even so, insert size will vary considerably (the vectors described here accept fragments rang-ing from 9 kb to 22 kb) so that a representative library constructed in this manner will be larger than the theoretical minimum. It should also be noted

High salt   Low salt

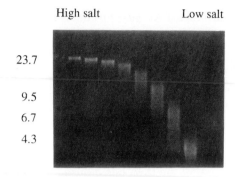

23.7

9.5

6.7

4.3

Figure 3. Analysis of donor DNA partially cleaved with *Sau*3A and fractionated by velocity gradient sedimentation. Samples of every third fraction were separated by electrophoresis in a 0.3% agarose gel as described in the legend to *Figure 2*.

that potentially unstable phage genomes just at the upper and lower limits of the packageable size range will be produced.

3.3 Vector structure

The structure of a replacement vector genome is illustrated in the context of EMBL3 (*Figure 4*; ref. 13). A stuffer fragment is flanked by a left arm encoding virion proteins and a right arm containing the major promoters, replication functions, and a few other essential genes. The DNA is shown in the linear form as isolated from phage particles. Each arm has a short single-stranded projection of 12 bases. These *cohesive ends* permit circularization following infection, and are regenerated during packaging. Although many replacement vectors follow the same overall design, members of the EMBL3*cos* series differ somewhat (15, 16). All of their essential genes are contained in a long right arm (*Figure 8*), an arrangement that expedites restriction mapping by partial cleavage methods (Section 10).

The stuffer fragment of EMBL3 is flanked by identical polylinkers in inverse orientation, each less than 40 bp in length:

*Sal*I–*Bam*HI–*Eco*RI—stuffer—*Eco*RI–*Bam*HI–*Sal*I

The polylinkers contain the only sites in the vector for the indicated enzymes (except for two additional *Sal*I sites in the stuffer fragment). Particularly important are the sites for *Bam*HI, an enzyme that makes a staggered break at the hexanucleotide sequence:

▼
GGATCC
CCTAGG
▲

*Bam*HI sites occur, on average, only once every 4^6 (4096) base pairs in random sequence DNA, rather too seldom for the 'random' fractionation of

48

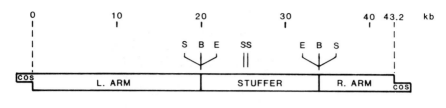

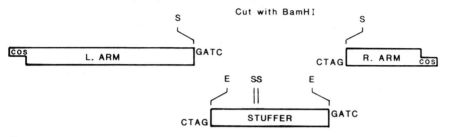

Figure 4. The EMBL3 genome. The stuffer fragment is flanked by polylinkers in inverse orientation. With the exception of two *Sal*I sites in the stuffer, the polylinkers contain the only sites for *Sal*I (S), *Bam*HI (B), and *Eco*RI (E). EMBL3 may thus be used to clone DNA fragments with termini generated by any of these enzymes. The single-stranded projections generated by *Bam*HI cleavage (GATC) also allow ligation of the arms to donor DNA fragments with termini generated by partial cleavage with *Sau*3A or *Mbo*I. *Sal*I sites retained in the arms permit the inserts of recombinant DNA molecules to be excised with minimum contamination by vector DNA sequences. *Eco*RI sites retained in the central fragment allow it to be enzymatically isolated. cos denotes a lambda cohesive end.

donor DNA. The core of the *Bam*HI site, however, is the recognition site for the enzymes *Sau*3A and *Mbo*I, which cleave it as shown:

$$\overset{\blacktriangledown}{} \text{GATC}$$
$$\text{CTAG} \overset{}{\blacktriangle}$$

*Bam*HI cleaved vector DNA can thus accept a donor DNA fragment prepared as described in Section 3.2. Inserts become flanked by *Sal*I sites, enabling their clean excision from the vector arms and their use as probes to recover overlapping recombinants from the library (Section 11).

Though the details vary, similar considerations apply to the other vectors discussed here (*Figure 8*). All contain *Bam*HI sites closely flanked by sites for other common restriction enzymes. Where flanking sites are unique to the polylinkers they can be used for cloning donor DNA cleaved with the appropriate enzyme, for example *Eco*RI and *Sal*I in the case of EMBL3. This may be useful if a desired sequence is known to be contained within an *Eco*RI or *Sal*I fragment that lies in the acceptable size range (9–22 kb). *Xho*I or *Sal*I sites also enable use of the partial fill-in cloning strategy described in Section 3.2.

In addition to cleavage sites, several replacement vectors have promoters for phage-encoded RNA polymerases (e.g. T3, T7, and SP6 polymerases) within their polylinkers. These allow the generation of RNA probes specific for either end of an insert, greatly facilitating chromosome walking (Section 11). The vector EMBL3cosW (16) has transcription terminators in its polylinkers in order to insulate phage genes from transcriptional interference by cloned DNA sequences.

3.4 Vector DNA preparation

The quality of vector DNA, while somewhat easier to control, is no less important than that of the donor DNA. Again it is important not to be too ambitious. Excessive starting material will produce a concentrated and thus viscous solution that is difficult to work with and that can be refractory to enzymatic cleavage, even after extensive phenol extraction. The purest phage are obtained by banding in equilibrium CsCl gradients. After removal of CsCl by dialysis the phage should be diluted appropriately, treated with proteinase K, and gently extracted several times with a phenol/chloroform/isoamyl alchohol mixture (13). Residual phenol can be removed by extensive dialysis, or by ethanol precipitation. Great care should be taken to maintain sterility of the vector DNA since integrity of the vulnerable cohesive ends is essential for infectivity.

When growing vector for DNA preparation use a single plaque (*Protocol 4*) to prepare a plate lysate (*Protocol 5*). Use the latter to infect a large scale bacterial culture (*Protocol 6*), purify the amplified phage (*Protocol 7*), and isolate vector DNA as described in *Protocol 8*.

Protocol 4. Growth of lambda as plaques

Equipment and reagents

- Sterile T broth: 1% tryptone, 1% NaCl
- Sterile 20% maltose (make with distilled water, filter through a 0.2 μm filter, and store at 4°C)
- Sterile 10 mM MgSO$_4$
- Sterile 250 ml conical flask
- Appropriate E. coli host (see Section 15)
- Incubator shaker at 37°C
- Sterile 50 ml polypropylene conical centrifuge tube
- Sterile phage buffer: 50 mM Tris–HCl pH 7.5, 100 mM NaCl, 10 mM MgSO$_4$, 0.1% gelatin
- Sterile CY top agarose: 1% casamino acids, 0.5% yeast extract, 0.3% NaCl, 0.2% KCl, 0.2% MgCl$_2$.6H$_2$O, 0.7% electrophoresis grade agarose
- Sterile glass tubes (e.g. 120 mm × 13 mm)
- 90 mm sterile disposable Petri dishes containing CY bottom agar: 1% casamino acids, 0.5% yeast extract, 0.3% NaCl, 0.2% KCl, 0.2% MgCl$_2$.6H$_2$O, 1.5% agar
- Water-bath or incubator at 45°C
- MSE Mistral 6L centrifuge (or equivalent)

A. *Preparation of plating cells*

1. Put 50 ml T broth in a sterile 250 ml conical flask, add 1 ml 20% maltose, and inoculate with the appropriate E. coli strain.

50

2. Incubate overnight at 37 °C in a rotary shaker, transfer to a 50 ml centrifuge tube, and pellet cells at 2500 r.p.m. for 10 min at room temperature in the 6L centrifuge.

3. Discard the supernatant and resuspend the pellet in 20 ml 10 mM $MgSO_4$. Although plating cells can be stored at 4 °C for several days, maximum recovery of *in vitro* packaged phage is likely to be achieved with freshly prepared cells. Rec⁻ cells (Section 15) should always be used fresh.

B. *Plating of phage*

1. Prepare serial tenfold dilutions of bacteriophage stocks or *in vitro* packaged phage (Section 4.2) in phage buffer.

2. Dispense 50 μl of each dilution to be assayed into a sterile glass tube, and add 50 μl plating cells.

3. Incubate at room temperature for 20 min to allow phage to adsorb to bacteria.

4. Add 3 ml of molten CY top agarose (45 °C), mix gently, and pour immediately on to the centre of a 90 mm Petri dish containing 25 ml of hardened CY bottom agar. Distribute the top agarose evenly by gentle swirling of plate. Repeat for all the other tubes of phage and bacteria.

5. Leave the plates to stand at room temperature for 5 min to allow the top agarose to set. Invert and incubate overnight at 37 °C.

6. Counting the number of plaques at a particular dilution enables one to calculate the number of plaque-forming units (p.f.u.) in the original stock.

Protocol 5. Preparation of phage stocks

Equipment and reagents

- Sterile phage buffer (*Protocol 4*)
- Chloroform
- Sterile micro- or Pasteur pipettes
- Sterile 1.5 ml microcentrifuge tubes
- Sterile glass tubes (*Protocol 4*)
- Plating cells (*Protocol 4*)
- Sterile CY top agarose (*Protocol 4*)

- 90 mm sterile disposable Petri dishes containing CY bottom agarose (*Protocol 4*)
- Sterile screw-capped 15 ml polypropylene centrifuge tubes (Falcon 15 ml conical centrifuge tubes or equivalent)
- Dimethylsulfoxide
- MSE Mistral 6L centrifuge (or equivalent)

A. *Picking plaques*

1. Using a Pasteur or micropipette, stab through a plaque into the bottom agar beneath.

Protocol 5. *Continued*

2. Eject agar core containing the plaque into 1 ml phage buffer containing 10 μl chloroform. Leave to stand for 1–2 h at room temperature, or overnight at 4 °C. The average plaque yields ≈ 10^6 phage particles.

B. *Preparation of plate lysate stock*

1. Mix 100 μl of eluted phage and 50 μl plating cells in a sterile glass tube. Incubate at 37 °C for 20 min.

2. Add 3 ml of molten top CY agarose (45 °C), mix, and pour immediately on to a 90 mm plate containing fresh CY bottom agar.

3. Next day the bacterial lawn should be completely lysed, with no individual plaques being visible. Add 5 ml of phage buffer to the plate and incubate at room temperature for ≥ 2 h.

4. After tilting the plate, harvest the phage suspension using a Pasteur pipette. Transfer it to a sterile polypropylene centrifuge tube.

5. Add 100 μl chloroform, mix by vortexing, and spin at 2500 r.p.m. for 10 min at 4 °C in the 6L centrifuge.

6. Transfer supernatant to a fresh tube, add a drop of chloroform, and store at 4 °C. The titre of the stock should be between 10^{10} and 10^{11} p.f.u./ml. Phage stocks can be stored at −70 °C after adding dimethylsulfoxide to 7%.

Protocol 6. Large scale preparation of phage

Equipment and reagents

- 500 ml conical flask containing 100 ml L broth: 1% tryptone, 0.5% yeast extract, 1% NaCl, 0.2% $MgCl_2.6H_2O$
- Four 2.5 litre conical flasks each containing 500 ml L broth
- Sterile phage buffer (*Protocol 4*)
- Phage stock (*Protocol 5*)
- Sterile 15 ml screw-capped polypropylene centrifuge tubes (*Protocol 5*)
- Chloroform
- Incubator shaker at 37 °C
- MSE Mistral 6L centrifuge (or equivalent)

Method

1. Inoculate 100 ml of L broth in a 500 ml conical flask with the appropriate bacterial host. Incubate overnight at 37 °C with shaking (250 r.p.m.).

2. Measure the OD_{600} of the culture and calculate the cell density assuming that 1 $OD_{600} = 8 \times 10^8$ cells/ml.

3. Withdraw four samples, each containing 10^{10} cells. Pellet at 2500 r.p.m. for 10 min at room temperature in the 6L centrifuge.

4. Discard supernatants and resuspend each pellet in 3 ml of phage buffer.

5. Add 5×10^7 p.f.u. to each resuspended pellet. Incubate for 20 min at 37 °C with occasional mixing. This number of p.f.u. works well for the EMBL series of vectors. For other lambda strains, the number may have to be determined empirically.

6. Add each infected sample to 500 ml pre-warmed L broth in a 2 litre flask. Incubate overnight at 37 °C.

7. Successful lysis generates a large amount of bacterial debris, which can be seen easily when the culture is held up to the light (the debris can vary in appearance from large clumps to a fine precipitate).

8. Add 10 ml of chloroform to each flask and continue shaking at 37 °C for 10 min. The lysate is stable at 4 °C.

9. Phage particles are purified from the lysate as detailed in *Protocol 7*.

Protocol 7. Purification of phage particles

Equipment and reagents

- DNase I and RNase A
- Solid sodium chloride, polyethylene glycol 8000, and caesium chloride
- Sterile phage buffer (*Protocol 4*)
- Chloroform
- 250 ml centrifuge bottles
- 50 ml polypropylene centrifuge tubes (*Protocol 5*)

- MSE Mistral 6L centrifuge (or equivalent)
- Sorvall RC5B Superspeed centrifuge, GSA rotor (or their equivalents)
- Beckman L8 ultracentrifuge, 50 Ti rotor, and 13.5 ml Quick-Seal polyallomer tubes (or their equivalents)

Method

1. To the 2 litres of lysate prepared as described in *Protocol 6*, add DNase I and RNase A to a final concentration of 1 µg/ml. Incubate for 30 min at room temperature.

2. Dissolve 117 g solid sodium chloride by swirling. Stand on ice for 1 h.

3. Transfer to 250 ml centrifuge bottles and pellet bacterial debris at 10 000 r.p.m. for 10 min at 4 °C in the RC5B centrifuge. Pool supernatants in a clean flask.

4. Dissolve 200 g solid polyethylene glycol 8000 by slow stirring with a magnetic stirrer. Stand in ice water for at least 1 h.

5. Recover precipitated phage by centrifugation as described in step **3**. Discard supernatant and drain pellets well.

6. Gently resuspend pellets (16 ml phage buffer per litre of supernatant in step **3**). Transfer to two 50 ml centrifuge tubes.

7. Add an equal volume of chloroform and vortex for 30 sec. Separate phases at 2500 r.p.m. for 10 min at room temperature in the 6L centrifuge.

Protocol 7. *Continued*

8. Remove aqueous phase suspension and add 0.75 g solid caesium chloride/ml. Mix gently to dissolve.

9. Transfer phage suspension to four 13.5 ml Quick-Seal tubes and top up with phage buffer containing 0.75 g/ml caesium chloride.

10. Spin at 38 000 r.p.m. for 24 h at 4 °C in L8 ultracentrifuge.

11. A bluish band of phage particles should be visible in the centre of the gradient. Puncture side of tube with a 21 gauge needle and withdraw phage with a 1 ml syringe. Puncture top of tube first to allow air to enter as caesium chloride solution is withdrawn.

12. Phage can be stored indefinitely at 4 °C in caesium chloride.

Protocol 8. Extraction of phage DNA

Equipment and reagents

- Dialysis buffer: 10 mM NaCl, 50 mM Tris–HCl pH 8.0, 10 mM MgCl$_2$
- Dialysis tubing. To prepare, boil pieces of convenient length for 10 min in a large volume of 2% sodium bicarbonate, 1 mM EDTA pH 8.0. Rinse thoroughly with distilled water and boil for 10 min in 1 mM EDTA pH 8.0. Cool and store at 4 °C. Handle with gloves.
- 0.5 M EDTA pH 8.0
- Proteinase K
- 10% SDS
- Phenol (*Protocol 1*)
- Chloroform/isoamyl alcohol (24 : 1)
- 3 M sodium acetate pH 7.0
- Ethanol
- TE (*Protocol 1*)
- 15 ml conical polypropylene centrifuge tubes
- MSE Mistral 6L centrifuge (or equivalent)

Method

1. Remove caesium chloride from phage purified as in *Protocol 8*. Dialyse against 1000-fold excess of buffer at room temperature for 1 h.

2. Replace buffer and dialyse for 1 h.

3. Transfer phage suspension to a 15 ml centrifuge tube. Add 0.5 M EDTA to final concentration of 20 mM.

4. Add proteinase K to 100 µg/ml and 10% SDS to 0.5%. Mix.

5. Incubate for 1 h at 50 °C.

6. Cool to room temperature and add an equal volume of phenol. Mix by gentle inversion. Add an equal volume of chloroform/isoamyl alcohol and mix well.

7. Separate phases at 2500 r.p.m. for 10 min at room temperature in the 6L centrifuge. Transfer aqueous phase to a clean tube.

8. Repeat steps **6** and **7**.

9. Extract aqueous phase once with chloroform/isoamyl alcohol.

10. Add sodium acetate pH 7.0 to 0.3 M, mix well, and add 2 vol. of ethanol. Recover DNA and estimate its concentration as described in *Protocol 1*, steps 11–14.

The DNA can be tested in several ways before use in cloning. Particular attention should be paid to the following points:

(a) The vector should have the expected physical and biological properties. The physical structure of the genome can be verified by restriction enzyme analysis, the biological properties of the phage by plating on appropriate indicator strains. Growth of red^+ gam^+ phage should be severely retarded on *E. coli* lysogenic for phage P2 (Section 13).

(b) *In vitro* packaging of uncleaved vector DNA should generate phage at the expected efficiency.

(c) Cleavage with *Bam*HI should dramatically reduce efficiency ($\geq 10^4$-fold).

(d) Ligation of cleaved vector DNA to itself should restore phage yield to within two orders of magnitude of that obtained with uncleaved DNA. Be alert to the possibility of damage to cohesive termini. Although quality has generally improved, it is not unknown for restriction enzymes to carry unadvertised nuclease activities.

3.5 Removal of the stuffer fragment

Removing the stuffer fragment, or otherwise preventing it from re-ligating to vector arms, maximizes the proportion of recombinant genomes in a library. Physical separation can be achieved by sedimentation through sodium chloride or sucrose gradients (1, 3), or by elution or extraction from an agarose gel (11, 12). The same qualifications apply as in application of these techniques to donor DNA (Section 3.2). Annealing the cohesive ends prior to separation (incubate at 42 °C for 1 h in 10 mM MgCl$_2$) allows both arms to be isolated as a single large DNA species, aiding resolution from the stuffer fragment.

An alternative to removal of the stuffer fragment is to render it incapable of ligation to *Bam*HI cleaved arms by cleavage with a second restriction enzyme. In the case of EMBL3, *Bam*HI sites are closely flanked internally by *Eco*RI sites (*Figure 4*). Cleavage with both enzymes releases short (< 10 bp) *Bam*HI–*Eco*RI fragments from both ends of the stuffer fragment, which can be removed by selective precipitation (13). This method is obviously limited to vectors in which appropriate restriction sites are juxtaposed in each of the two polylinkers. Since enzymes differ in the efficiency with which they cut terminal sites (see New England Biolabs catalogue and ref. 17), it is generally preferable to cleave sequentially leaving the better enzyme till last. In the absence of specific information, determine an order of cleavage by trial and error, assaying on the basis of phage titre following ligation and packaging in the presence and absence of donor DNA.

An even simpler and very effective means of preventing re-assimilation of the stuffer fragment is to treat cleaved vector DNA with alkaline phosphatase. This can reduce non-recombinant background virtually to zero (15), but necessarily precludes phosphatase treatment of donor DNA. A method for the preparation of dephosphorylated vector DNA sufficient for the production of several libraries is given in *Protocol 9*.

Protocol 9. Preparation of dephosphorylated vector DNA

Equipment and reagents

- Caesium chloride purified DNA (*Protocols 7* and *8*)
- Appropriate restriction enzyme (*Bam*HI for cloning *Mbo*I or *Sau*3AI partial digest fragments), and 10 × digest buffer
- BSA (10 mg/ml)
- Spermidine (0.1 M made up in sterile distilled water, filtered through a 0.2 μm filter, and stored at −20 °C)
- 0.5% agarose mini-gel
- Calf intestinal alkaline phosphatase (Boeh-

- ringer-Mannheim; molecular biology grade)
- 0.5 M EGTA pH 8.0
- Phenol (*Protocol 1*)
- Chloroform/isoamyl alcohol (24:1)
- 5 M ammonium acetate
- Ethanol
- Microcentrifuge and 1.5 ml microcentrifuge tubes
- TE (*Protocol 1*)

Method

1. Digest 20 μg vector DNA with a two- to threefold excess of *Bam*HI in a volume of 100 μg for 60 min at 37 °C. As well as 1 × digest buffer, the reaction should contain 100 μg/ml BSA and 5 mM spermidine.

2. Analyse 1 μl (0.2 μg) on a 0.5% agarose mini-gel. If digestion is incomplete, add more enzyme, and continue the reaction.

3. When digestion is complete, add 1 U of alkaline phosphatase and continue incubating at 37 °C for 30 min.

4. Add 0.1 vol. 0.5 M EGTA and incubate at 65 °C for 60 min.

5. Extract twice with an equal volume of phenol/chloroform/isoamyl alcohol (25:24:1), and once with an equal volume of chloroform/isoamyl alcohol. Separate phases by centrifugation in a microcentrifuge. Remove aqueous phase to a fresh tube after each extraction.

6. Add 1 vol. of 5 M ammonium acetate and 2 vol. of ethanol. Mix. Allow DNA to precipitate at room temperature for 10 min.

7. Recover DNA by centrifuging for 10 min in a microcentrifuge. Dry pellet briefly under vacuum, and resuspend at a concentration of 1 mg/ml in TE.

Points to note about the use of phosphatase are:

(a) Restriction endonuclease cleavage and dephosphorylation of vector DNA can be performed simultaneously under reaction conditions recommended for a given restriction enzyme. Add alkaline phosphatase

(Boehringer-Mannheim molecular biology grade; one unit per five pico-moles of ends) 30 minutes before the end of the cleavage. Spermidine (5 mM) assists cleavage as well as helping to maintain integrity of the cleaved ends.

(b) It is critical to inactivate/remove alkaline phosphatase before proceeding. This is achieved by adding EGTA pH 8.0 to 50 mM, incubating at 65 °C for 60 min, and then phenol extracting and ethanol precipitating the DNA.

(c) Efficiency of dephosphorylation can be checked by attempting to ligate phosphatase treated vector DNA to itself. Integrity of the dephosphoryl-ated ends can be checked by ligating phosphatase treated vector DNA in the presence of unphosphatased donor DNA fragments of appropriate length. These reactions can be monitored by agarose gel electrophoresis: dephosphorylated vector DNA should only show a shift in mobility when ligated to DNA that retains its 5′ phosphate group. A more precise estimate will be obtained by using an assay based on the ability to form packageable DNA molecules.

(d) A heat-labile, and thus easily inactivated, phosphatase is available (Epi-centre Technologies, Madison, USA).

Vector background can also be reduced by using the partial fill-in strategy discussed in Section 3.2.

4. Generation and recovery of recombinant DNA molecules

4.1 Ligation

Although there is a strong theoretical foundation for the calculation of vector and donor DNA concentrations that maximize the formation of a desired ligation product (1), empirical methods tend to be used more often. In thinking about ligation conditions it is simplest to consider the left and right vector arms as a single component (they readily anneal via their cohesive ends). It is also useful to distinguish between efficiency calculated per micro-gram of vector DNA and that calculated per microgram of donor DNA. One can also take into account the way in which ones vector DNA was prepared; whether or not dephosphorylated, whether or not depleted of stuffer frag-ment. Finally, it should be borne in mind that while vector DNA concentra-tion is likely to be known relatively precisely, donor DNA concentration will have been determined only roughly.

  The preferred substrate for *in vitro* packaging is a concatemer, an end to end chain in which recombinant genomes are flanked on both sides by a *cos* site (*Figure 7*). To favour concatemer formation, the total DNA concentration should exceed 100 μg/ml. Typically, 2 μg of cleaved and dephosphorylated vector is added to 0.5 μg of size selected donor DNA in a final volume of 10 μl. The mixture is incubated overnight at 12 °C in the presence of one unit of T4 ligase (15). Where the actual concentration of the participating DNA species is only roughly known, optimal conditions can be calculated empirically by ligating varying amounts of donor DNA to a constant amount of vector DNA. Trial ligations can also be used to find a donor DNA fraction containing fragments near to the maximum size for the vector.

Since $\geq 10^7$ recombinants per microgram of insert DNA can be recovered by the use of high efficiency packaging extracts (Section 4.2) and restriction deficient hosts (Section 15), sufficient clones for a full genomic library will often be generated in a test ligation, without the need for scaling up.

Note: it is important to verify the properties of host strains at this stage, particularly restriction and recombination status (Section 15).

4.2 Packaging

Greater than 10^9 infective phage per microgram of lambda DNA can be obtained by *in vitro* packaging. This efficiency allows more than 10^7 recombinant DNA molecules to be recovered per microgram of donor DNA. *In vitro* packaging is carried out in 'packaging extracts', which are prepared following the induction of *E. coli* lysogenic for certain mutant derivatives of lambda. The extracts contain all of the structural proteins necessary for the assembly of lambda DNA into an infective phage particle, and are preferably prepared from restriction deficient *E. coli* strains (Section 15.1). Packaging extracts are available commercially, can be stored at −70 °C until required, and usually work well following the protocols supplied by the manufacturer.

A description of the preparation of packaging extracts is beyond the scope of this article, and the availability of high efficiency extracts from commercial sources may in any case obviate the need for individual laboratories to prepare their own. If a lot of trial reactions are anticipated, however, or if commercial extracts are too expensive, home-made extracts may be worth the investment of time (1, 56). Although different methods of preparation yield extracts with different preferences with respect to genome size, most packaging extracts available commercially impose a fairly stringent selection for DNA molecules close in size to that of the wild-type lambda genome.

Note: packaging extracts deficient in the *A* gene product prefer monomers (probably cohered) to concatemers (see ref. 36).

4.3 Transfection

It is possible to generate a library representing a small (e.g. bacterial) genome by transfection, the uptake of naked phage DNA by bacteria (2). The desirable substrate for transfection is a linear monomeric phage DNA molecule. Optimum ligation conditions for transfection are thus not necessarily the same as those for *in vitro* packaging.

4.4 Scaling up

Having determined the optimum ligation conditions and taken background into account, set-up a large scale reaction that will generate at least the number of recombinant phage needed for a 'complete' library (Section 5), and package the ligated DNA in extracts of the highest efficiency.

Titre the packaged phage and store the library at 4°C until required. Addition of a drop of chloroform prevents bacterial growth. Chloroform also reduces plating efficiency, however, so samples of packaged material should be warmed to 37°C for ten minutes before they are mixed with plating cells. Packaging extracts can themselves inhibit bacterial growth if present in excess. This is only likely to present a problem if the library is of very low titre. Titre of phage packaged *in vitro* will drop with storage, so it is advisable to recover recombinant phage as plaques at the earliest opportunity.

Generating a library involves a series of reactions. Errors can occur, sometimes for trivial reasons, at a number of stages. Assuming that the error can be recognized as such, which will be the case if the appropriate controls have been included, it should be a simple matter to rescue the operation. It obviously helps if you have excess starting material, and have retained a proportion of the reaction mixture at intermediate stages.

4.5 Growth of phage as plaques and the preparation of stocks

Methods for propagation of lambda as plaques, and for preparation of phage stocks, can be found in *Protocols 4* and *5*. Points to note are:

(a) Phage remain infective for years if stored at 4°C, in the absence of detergent, in either phage buffer or broth containing 10 mM $MgSO_4$. Addition of a few drops of chloroform helps to maintain sterility. Dilute in phage buffer to titre (even a plaque contains $\approx 10^6$ phage).

(b) Efficient adsorption of phage requires the presence of magnesium ions, and *lamB*-encoded receptors in the host bacterial membrane. Since phage can adsorb to dead cells and debris, old cultures are inadvisable. Inclusion of maltose in the medium induces the maltose operon, which includes the *lamB* gene. In cases where bacterial hosts do not grow well in T broth, substitute L broth (*Protocol 6*).

(c) Freshly poured plates should not be used for plating lambda as droplets of moisture form on the surface of the top agarose during incubation at 37°C causing plaques to streak and run into one another. Plates should be dried either by leaving at room temperature for one to two days, or with the lids off in a laminar flow cabinet. Properly dried plates show a slight puckering of the agar surface. Plates can be stored at 4°C for several weeks. They should be allowed to warm to room temperature before use, however.

(d) After overnight incubation at 37 °C, plaques should be visible as clearings in an opaque background of unlysed bacteria. Plaques produced by different strains of lambda, or by different recombinants in the same vector background, can vary in size (0.1–2 mm). Adsorption of phage to bacterial cells *before* plating helps reduce heterogeneity of plaque size.

(e) Plaque size does not increase once bacteria have reached stationary phase. As a result, factors that increase the time to form a confluent lawn (e.g. less rich growth medium) tend to increase plaque size and vice versa. To restrain plaque size, use a rich medium such as L broth (*Protocol 6*) for making the top agarose and bottom agar. To increase plaque size (such as when plating *red– gam–* phage; see Section 13), use CY medium (*Protocol 1*), or BBL medium (1% BBL trypticase, 0.5% sodium chloride).

5. Library size

Were a genome of known size to be represented as a truly random collection of inserts of equal length, it would be a simple matter to determine the size of library for a given probability of including any particular sequence. In practice, even though several of these conditions do not strictly obtain, it is possible to arrive at an approximation sufficient for most purposes.

Until recently, the generally accepted means of calculating library size was as follows. If x is the insert size and y is the size of the haploid genome (in the same units), then a library of

$$N = \frac{\ln(1 - p)}{\ln(1 - x/y)}$$

clones should have a probability p of containing any particular DNA sequence (18). Assuming 15 kb inserts and making $p = 0.99$ we obtain the sizes shown in *Table 1* (upper figures). We also show the sizes of library calculated for 35 kb fragments cloned using a cosmid vector. It should be noted that non-haploid organisms, unless isogenic, will require larger libraries for complete representation.

Use of a different formula (19) suggests that library sizes estimated as above are likely to be underestimates by a factor of two to four (*Table 1*, lower figures). It is probably best to err on the safe side.

6. Limiting amounts of donor DNA

Provided that starting material is not limiting, one would be advised to begin with about 100 µg of high molecular weight donor DNA. 5–10 µg will be needed to determine the optimum partial cleavage conditions. A considerable proportion of the partially cleaved material will fall outside of the desirable

Table 1. The number (N) of independent recombinants for a 99% probability of having cloned a particular DNA sequence in a lambda vector (15 kb inserts) or a cosmid vector (35 kb inserts)

Organism	Approximate DNA content of haploid genome (bp)	N[a] for 15 kb inserts	N[a] for 35 kb inserts
Escherichia coli	4.2×10^6	1.3×10^3	5.5×10^2
(bacterium)		2.9×10^3	1.1×10^3
Saccharomyces cerevisiae	1.4×10^7	4.3×10^3	1.8×10^3
(bakers yeast)		1.1×10^4	4.2×10^3
Drosophila melanogaster	1.4×10^8	4.3×10^4	1.8×10^4
(fruit fly)		1.3×10^5	5.2×10^4
Stronglyocentrotus purpuratus	8.6×10^8	2.6×10^5	1.1×10^5
(sea urchin)		8.9×10^5	3.6×10^5
Homo sapiens	3.3×10^9	1.0×10^6	4.3×10^5
(man)		3.7×10^6	1.5×10^6
Triticum aestivum	1.7×10^{10}	5.2×10^6	2.2×10^6
(hexaploid wheat)		2.1×10^7	8.6×10^6

[a] The upper values were derived using an equation given in ref. 18. Although these are likely to be adequate for most purposes, ref. 19 suggests that two to four times as many clones should be screened to be sure of complete representation (lower values).

size range, and there may be losses at various stages of the procedure, for example during dialysis or ethanol precipitation of salt gradient fractions.

It is possible to construct libraries on a much smaller scale, for example from DNA isolated from a non-renewable source such as a small tumour. Clearly extreme care should be taken in preparation of the DNA and, to avoid wastage, conditions for partial cleavage could be determined on a less important but similarly prepared sample. Alternatively, a *Dam* methylase/*Mbo*I partial cleavage strategy can be used (20). Partial fill-in and dephosphorylation of the vector DNA, and partial fill-in of the donor DNA, will serve to maximize yield of recombinant clones. Such a combination of strategies has allowed the generation of 'sub-libraries' from nanogram quantities of gel-purified YAC DNA (21).

7. Amplification

Assuming the growth characteristics of recombinants to be unaffected by sequence content, and an appropriate choice of vector and host, all regions of the genome will be proportionally represented as viable phage. If some recombinants are more vigorous than the norm this may be reflected in plaque size, but the vast majority of recombinant DNA molecules will give rise to plaques, even the smallest of which can be detected by hybridization methods.

A purpose built library may be packaged, plated, screened, and discarded. Either the desired sequence will have been found, or it is not contained in the library. If necessary more DNA can be packaged and screening repeated. If high cloning efficiencies are being obtained routinely it might be argued that only purpose built libraries are necessary. Cleaved donor and vector DNAs can be stored and combined as required. On the other hand, it may be deemed prudent to produce a more permanent library at a time when everything is working and all the necessary ingredients are fresh and to hand.

Amplified libraries of high titre are easily produced. They are a useful resource for distribution to others and from which genes may be isolated over a period of time. Their major drawback is the inevitability of at least some differential growth leading, in the extreme, to effective loss of particular classes of recombinant. Proportional misrepresentation is exacerbated by competitive growth in liquid culture, or at very high plaque density. For this reason amplification should be by growth as plaques at relatively low density, and it is advisable to start with more than a 'complete' library.

To minimize heterogeneity of plaque size, use fresh plating cells to which the phage have been pre-adsorbed (see Section 4.5), and keep the plates absolutely horizontal until the top agar has set. After growth, overlay the agar surface with L broth or phage buffer and leave for more than one hour at room temperature, or overnight at 4 °C. Pipette the phage suspension into a centrifuge tube and spin at low speed to remove bacterial debris. Alternatively, scrape the top agar into a sterile flask containing phage buffer (*Protocol 4*) and 5% chloroform. Again, leave the phage to elute at room temperature for at least one hour (or overnight at 4 °C) with gentle mixing. Remove the agarose and debris by centrifugation at room temperature (2500 *g* for 10 min in sterile screw-capped 50 ml polypropylene centrifuge tubes). Store phage suspensions in capped tubes at 4 °C in the presence of 2% chloroform (to suppress bacterial growth). Titre lysates as described in *Protocol 4*. Re-amplification of such a library is inadvisable due to the potential for overgrowth by a subpopulation of relatively vigorous phage.

Methods for the long-term storage of libraries, either on plates or in the form of filter replicas, have been described. Such methods allow an *unamplified* library to be screened many times over a period of time (see Section 11).

8. Screening by hybridization

If all has been successful up to this point, the proud possessor of a genomic DNA library turns his or her mind to the problem of finding a particular sequence of interest from among as many as 10^6 independent clones. This would be like looking for the proverbial needle in a haystack were no probe or selection system available. For the purposes of this section we assume that screening is to be carried out by hybridization and that a suitable probe, be it

DNA or RNA, is available. Other screening methods are considered in Section 9.

Hybridization is carried out not on plaques themselves, but on nitrocellulose or nylon membranes to which the plaques have been blotted (1). Each plaque is a source both of amplified phage particles and a considerable amount of unpackaged phage DNA. An agar surface in a Petri dish or other receptacle (cafeteria trays have been used), on which a portion of the library has been plated in agarose top layer, is brought into contact with a sterile circle or sheet of membrane. Both phage particles and free DNA are transferred, and if the surface is not too wet spreading is minimal. When the membrane is removed it carries an invisible replica of the pattern of plaques. A number of replicas can be taken from each plate. Agarose instead of agar top layer discourages lifting when the membrane is removed, and may also reduce non-specific binding of the probe.

Replicas are treated briefly with strong alkali to release DNA from phage particles and to denature it. The DNA remains single-stranded when a more neutral pH is restored, and is irreversibly fixed to the membrane by baking at 80 °C (nitrocellulose and nylon) or by UV crosslinking (nylon only). The membrane is 'pre-hybridized' to saturate its non-specific DNA binding capacity, and probed with a labelled single-stranded DNA (or RNA) species. Under conditions of temperature and salt concentration that favour the renaturation of homologous, or partially homologous, nucleic acid species some proportion of the probe will become bound to regions of the filter where there are related DNA sequences. Authentic binding is resistant to washing in probe-free solutions, and can be visualized by autoradiography.

It is a simple matter to match up the 'positive signals' with the corresponding plaques (or regions) on the agar plate. Authentic signals appear in equivalent positions on both of a pair of duplicate filters, distinguishing them from artefactual signals.

1000 or more small plaques can be resolved on the surface of a 90 mm diameter plate. In the case of an *E. coli*- or yeast-sized library, only 10^3–10^4 independent recombinants will be required for 'complete' representation ($p = 0.99$). Thus one to ten 90 mm plates can carry a complete library. With luck a positive signal will be found to correspond with a fairly well separated plaque, which can be picked independently of its neighbours. Low density screening may even be feasible for a *Drosophila*-sized genome (3×10^4 recombinants for $p = 0.99$), but is inconvenient for mammalian or plant genomes (10^6 to 5×10^6 recombinants necessary). It is thus normal to screen large libraries at a much higher density (e.g. 2×10^5 per 220 × 220 mm plate). At this density, lysis is almost confluent across the surface of the plate and it is impossible to identify and pick out an individual plaque. It is therefore necessary to carry out further rounds of screening. This involves taking a plug of agar from the region corresponding to the positive signal and soaking it in phage buffer to elute the phage particles. Dilutions of the phage suspension are plated out at a density of 500–1000 plaques per 90 mm plate, and rescreened. It is usually necessary to carry out yet another round of screening to be confident of having a single clone.

Phage can diffuse across the surface of an agar plate, especially if it has been blotted. It is thus advisable to re-purify a positive plaque by a further round of dilution and plating. Alternatively, single plaques may be obtained by streaking (1). To be on the safe side, pick at least two well-separated plaques, and grow them on a small or large scale for DNA isolation and characterization.

A method for the screening of a mammalian genomic library (10^6 recombinants) is given in *Protocol 10*.

Protocol 10. Screening a lambda library with a radiolabelled probe

Equipment and reagents

- 250 ml of CY top agarose and 2.5 litres of CY bottom agar (*Protocol 4*)
- Nunc bioassay dishes (Gibco-BRL; 245 × 245 mm)
- Nitrocellulose or nylon filters (220 × 220 mm; Schleicher and Schuell BA85 or Hybond N); nylon is recommended when repeated screening of a filter set is contemplated
- 50 ml sterile screw-capped polypropylene tubes (*Protocol 1*)
- Plating cells (*Protocol 4*)
- Whatman 3 MM filter paper (or equivalent)
- 21 gauge syringe needles and waterproof drawing ink
- 0.5 M NaOH, 1.5 M NaCl
- 0.5 M Tris–HCl pH 7.0, 1.5 M NaCl

- 20 × SSC: 3.0 M NaCl, 0.3 M sodium citrate, adjust pH to 7.0
- UV transilluminator or oven at 80°C
- Polyethylene freezer box (240 × 240 mm)
- Water-bath at 65°C
- 50 × Denhardt's solution: 1% Ficoll, 1% polyvinylpyrrolidone, 1% bovine serum albumin (sterilize by filtration through a 0.45 μm filter and store at −20°C)
- 10% SDS
- 10% sodium pyrophosphate
- Radioactive ink (made by mixing a small amount of ^{32}P with waterproof drawing ink)
- X-ray film (30 × 40 cm)
- X-ray cassettes with intensifier screens (30 × 40 cm)

Method

1. Prepare five Nunc bioassay dishes of CY bottom agar by pouring 250 ml of molten agar (cooled to 45°C) into each dish. Remove bubbles by passing a Bunsen flame over surface before agar hardens. Dry plates either by leaving them uncovered under a laminar flow hood with the air blower on for 1–2 h, or by leaving them with their lids on at room temperature for two days.

2. To 1 ml plating cells in each of five 50 ml screw-capped tubes, add 2 × 10^5 recombinant phage in a volume of 1 ml or less. Incubate at 37°C for 20 min.

3. Add 45 ml of CY top agarose (45°C) to each tube of infected cells, and pour on to the surface of a Nunc plate pre-warmed to 37°C. Spread agarose quickly across surface of bottom agar by tilting plate. After agarose has set, incubate the plates inverted at 37°C overnight.

---

4. Cool plates at 4°C for 60 min to harden top agarose.

5. Number ten 220 × 220 mm nitrocellulose or nylon filters with a soft lead pencil or ballpoint pen. Two filters are needed per plate. Use directly from the box with no pre-treatment (i.e. no pre-wetting or sterilization). To ease placement and removal of filters during blotting, trim them to 210 × 220 mm.

6. To make the first replica of each plate, place a filter on to the surface of the top agarose. Air bubbles between filter and top agarose can be avoided by bending filter into a U-shape so that its middle contacts the centre of the plate first. The rest of filter can then be lowered into position. Do not move the filter once contact with the plate is made. The filter is keyed to the plate by stabbing through the filter and into the agar beneath using a syringe needle dipped in drawing ink. Make an asymmetric pattern of holes.

7. After 1 min on plate, carefully peel away the filter and place it plaque side up for 5 min on a sheet of Whatman 3 MM paper soaked in 0.5 M NaOH, 1.5 M NaCl.

8. Transfer filter to another sheet soaked with 0.5 M Tris–HCl pH 7.0, 1.5 M NaCl. Leave for 5 min.

9. Rinse filter by placing for 5 min in a tray containing 1 litre 2 × SSC. If any top agarose has adhered to the filter, it can be removed by gentle rubbing with a tissue soaked in 2 × SSC. Dry the filter at room temperature on a sheet of 3 MM paper.

10. The second replica is made exactly as described in steps **6–9**, except that the filter is left in contact with the top agarose for 2 min. Stab through the filter with a syringe needle in the positions indicated by the ink spots/holes in the agar beneath.

11. Once all blots have been prepared and dried at room temperature, fix DNA to them by baking at 80°C for 2 h (sandwiched between 3 MM paper). Alternatively (nylon filters only), UV irradiate on a trans-illuminator for 2–5 min. Over-baking of nitrocellulose causes it to become very brittle.

12. Wet the blots with 2 × SSC, place them in a 240 × 240 mm polyethylene freezer box, and add 150 ml of 5 × SSC, 5 × Denhardt's solution, 0.1% sodium dodecyl sulfate (SDS), 0.1% sodium pyrophosphate. De-natured sonicated salmon sperm DNA can be added to 100 μg/ml if background is a problem, but is not usually necessary. Incubate in a water-bath at 65°C for 1–4 h *without* shaking. Replace with 100 ml new hybridization mix and 50 ng radiolabelled probe. Incubate over-night at 65°C.

13. Dispose of the hybridization mix safely, add 500 ml 2 × SSC, 0.1% SDS, 0.1% sodium pyrophosphate (pre-heated to 65°C), and incubate

Protocol 10. *Continued*

with shaking in a 65 °C water-bath for 15 min. Wash twice more for 15 min with 500 ml of 0.1 × SSC, 0.1% SDS, 0.1% sodium pyrophosphate (pre-heated to 65 °C). These hybridization and washing conditions are for perfectly matched single copy probes. Less stringent conditions may be necessary for other types of probe.

14. Seal filters inside plastic bags while still damp. Autoradiograph with intensification at −70 °C. Positive signals are normally detectable after overnight exposure. To facilitate alignment of filters, backing paper (e.g. Whatman Benchkote) marked with several asymmetrically spaced dots of radioactive ink should be placed in the autoradiography cassettes.

15. After exposure and development, align the autoradiograph with the backing paper. Mark the positions of the keying holes on the filters. Trace the positions of the keying holes and any positive hybridization signals on to an acetate. Bona fide signals should appear in the same position on each of the two blots corresponding to a particular plate. Identify the area of the plate corresponding to the positive signal by aligning the keying holes on the plate with the marks on the acetate.

16. Stab through the relevant region of the agar using the wide end of a yellow micropipette tip. Expel the core into 5 ml of phage buffer containing 50 μl chloroform. Elute phage by incubation at room temperature for 2 h or longer. For secondary screening, dilute 10 μl of eluted phage in 1 ml phage buffer. Plate out 10 μl as described in *Protocol 5*.

Points to note when screening a library are:

(a) Plating at near confluence encourages simultaneous infection. Unless a Rec⁻ strain is used (Section 15.2), this could lead to genetic recombination between repetitive elements present on different DNA molecules. The result might be viable phage that carry sequences derived from two different regions of the genome. If high density screening on Rec⁺ hosts is unavoidable, the potential for rare juxtaposition should be remembered when the recombinant DNA molecules are analysed.

(b) In addition to plaque-shaped spots, non-specific 'hybridization' may be encountered despite exhaustive washing, e.g. blotchy or uniform distribution of radioactivity across the membrane, or significant labelling of all plaques (only distinguishable as such in low density screening procedures). Heterologous probes, or complex mixtures of radioactive cDNA or messenger RNA, may be especially problematic. Unfortunately, the very recombinant phage that one is trying to isolate is often the only true control for both signal-to-noise problems and, ultimately, the presence or absence of the desired sequence in the library.

(c) In many cases a probe will be a DNA fragment cloned in a plasmid vector. A number of plasmid vectors in current use, notably cosmids and certain expression vectors, cross-hybridize with lambda DNA.

(d) Primary libraries plated in glycerol top layer can be stored for long periods, either on the plates or on filter replicas. This technique makes repeated screening of a primary library over time a quite straightforward matter (Section 11).

The suggested hybridization conditions work well with the recommended filters. Different conditions may work better with different brands. Use the manufacturers' recommendation.

9. Screening by recombination *in vivo*

Plaque screening by hybridization is by far the most popular means of identifying phage with inserts related to a nucleic acid probe. An interesting alternative, however, is *in vivo* selection for phage with the desired characteristics (22). A library is used to infect *E. coli* in which the probe sequence is present as a component of a plasmid replicon. Where homologous insert sequences allow, integration of the plasmid into a phage genome can occur by recombination. Phage genomes containing sequences of interest are thus tagged by whatever marker is carried by the plasmid and, provided that the hybrid genome is not too long to be packaged, can be recovered as phage particles.

As originally described, selection relied upon a vector with two amber mutations, and a probe plasmid that encodes a suppressor tRNA. Propagation of the library on plasmid-containing, recombination proficient, *E. coli* allows both growth of the phage and homology-dependent integration of the plasmid into the phage genome. Re-plating on a suppressor-free host selects for phage with the desired characteristics.

In vivo selection is potentially a rapid, inexpensive, and sensitive alternative to plaque hybridization. It has been dogged, however, by a high background of phage that escape dependence on suppressor, and by low recombination frequency. The former problem is largely due to occasional loss of amber mutations by genetic exchange with defective phage from packaging extracts. It can be side-stepped by a strategy involving suppression of amber mutations in the host rather than the vector (23). One mutation is in *dnaB*, a bacterial gene required for phage DNA replication (though both phage and host DNA replication are normally *dnaB*-dependent, a substitute can be provided for host replication), the second is in *lacZ*. Blue plaques in the presence of Xgal confirm presence of the suppressor. Efficient recombination between plasmids and lambda genomes requires the product of the lambda gene *rap*. This gene is missing from most lambda replacement vectors but can be provided on an ancillary plasmid (24).

Before a selected recombinant is analysed or further manipulated it is desirable that the plasmid be shed cleanly by an excisive recombination event. White plaques on an appropriate *lacZ* indicator will reveal phage that have lost the plasmid.

10. Analysis of recombinant DNA molecules

Having identified one or more recombinant phage with the appropriate characteristics, it will be necessary to analyse their genomes in order to determine their exact relationships with the probe and with each other, and to develop subcloning strategies suitable for further characterization and/or manipulation. A method for the large scale preparation of lambda DNA is given in *Protocol 6*. Analysis of individual recombinants (e.g. restriction mapping and probe preparation for chromosome walking) can usually be carried out using DNA prepared on a small scale, however. Although a number of mini-preparation procedures have been described, very few produce DNA of sufficient purity. The method detailed in *Protocol 11* (a modification of the method described in ref. 55) produces DNA that is a good substrate for restriction enzyme digestion, riboprobe preparation, and DNA sequencing.

Protocol 11. Mini-preparation of lambda DNA

Equipment and reagents

- 1 ml of plate lysate prepared as described in *Protocol 5B*. The optimal input of phage for a plate lysate depends on the characteristics of the phage and host. For EMBL derivatives grown on a *recD⁻* host, use ≈ 10^5 p.f.u.
- DE52 ion-exchange cellulose (Whatman pre-swollen DEAE–cellulose). Prepare according to manufacturer's instructions. Then equilibrate in 10 mM Tris–HCl pH 7.0, 1 mM EDTA, with 0.01% sodium azide as a preservative. Store at 4°C as a 40% slurry. Before use, wash with L broth (*Protocol 6*) twice, and resuspend as an 80% slurry in L broth.
- DNase I (10 mg/ml; made fresh in sterile distilled water)
- Microcentrifuge and 1.5 ml microcentrifuge tubes
- 0.1 M EDTA pH 7.0
- Pronase (20 mg/ml in sterile distilled water)
- 5% cetyl trimethyl ammonium bromide (CTAB) made up in 0.5 M NaCl
- Heating block or water-bath at 68°C
- 1.2 M NaCl
- Ethanol
- 70% ethanol
- TE (*Protocol 1*)

Method

1. To 1 ml of plate lysate, add 2 μl DNase I and incubate for 30 min at room temperature.

2. Centrifuge lysate at full speed in a microcentrifuge for 10 min. Transfer supernatant to a fresh tube.

3. Add 400 μl 80% DE52/L broth and mix for 10 min at room temperature on an angled rotating wheel, or horizontal shaker. Centrifuge for 2 min in a microcentrifuge. Transfer supernatant to a fresh tube.

4. Add 250 μl 0.1 M EDTA pH 7.5 and 4 μl pronase. Incubate at 45 °C for 15 min.

5. Add 25 μl of CTAB and incubate at 68 °C for 3 min. Cool on ice for 5 min. Centrifuge in a microcentrifuge for 10 min.

6. Carefully remove supernatant and discard. Resuspend pellet in 200 μl 1.2 M NaCl and add 500 μl ethanol. Mix and leave at room temperature for 10 min. Centrifuge for 10 min in a microcentrifuge.

7. Carefully remove supernatant and discard. Wash pellet with 500 μl 70% ethanol, vacuum dry, and resuspend DNA in 20 μl of TE.

8. 2×10^{10} p.f.u. yields 0.75–1 μg DNA.

A preliminary step in characterization of a recombinant DNA molecule is often the construction of a 'restriction map' showing sites for enzymes that cleave relatively infrequently (usually enzymes with a hexanucleotide target). This task is made easier if an overlapping set of clones is available, since regions of overlap give clues to the relative positions of specific restriction fragments (*Figure 5*). Use of an enzyme that cleaves within the retained section of the polylinker (e.g. *Sal*I in the case of EMBL3: *Figure 4*) enables fragments from the extremities of inserts to be recognized. These, unlike fragments entirely contained within inserts, vary in size from molecule to molecule (because of their 'random' ends).

A more direct approach to mapping, better suited to the rapid generation of data during chromosome walking, involves the identification of partial cleavage products of recombinant genomes labelled at one end. Although several end-labelling protocols have been described (reviewed in ref. 16), the simplest and most widely applicable is the '*cos*-labelling' strategy (25). Here, the ends of a lambda genome are asymmetrically labelled using radioactive oligonucleotides that hybridize specifically with either the left or the right cohesive end. Asymmetrically labelled DNA is then partially cleaved with a restriction enzyme and a map is deduced from the ladder of partial cleavage fragments visualized after gel electrophoresis and autoradiography. Because only DNA fragments containing a labelled end can be visualized, successive rungs of the ladder correspond to cleavage at successive restriction sites. The vectors EMBL3*cos*, EMBL3*cosNot* (15), and EMBL3*cos*W (16) were designed specifically to expedite this strategy. All of the coding information normally present in the left and right arms has been placed in one long right arm, leaving the left cohesive end separated by only 200 bp, rather than 20 kb, from an insert. This offers significant advantages over conventional vectors in which the presence of long arms shifts fragment lengths into a range that is more difficult to measure accurately. Map construction is further simplified in the case of EMBL3*cosNot* and EMBL3*cos*W by the inclusion of a *Not*I site in the right hand polylinker. Since *Not*I cleaves very rarely (it has an octanucleotide target), this makes it likely that a map is made only with respect to sites within an insert (16).

The stuffer fragment of EMBL vectors and their derivatives shares homology with the right arm of the vector (13). This can give confusing results if recombinant DNA molecules are analysed by heteroduplexing with vector DNA (H. Delius, personal communication). A recombinant DNA molecule with an unrelated insert should be used instead.

11. Chromosome walking

Libraries containing an overlapping set of DNA fragments representative of the entire genome are ideally suited to chromosome 'walking', the systematic isolation of clones representing regions of the genome progressively more distant from a specific starting point (see *Figure 5*). At least in theory, any point on a chromosome can be reached by walking from any other point on the same chromosome. On a small scale, walking can allow cloning of an entire gene and its control sequences, starting only with a gene fragment or cDNA. On a larger scale, walking provides access to a genetic locus from a cloned entry point somewhere in its vicinity. In extreme cases, the only means of identifying a target gene may be an extensive analysis of clones that encompass the general region in which it is located. Walking has become an important tool for cloning determinants of human inherited disorders, a good example being the cystic fibrosis gene (26).

To maximize the ground covered during each step of a walk, it is important to be able to generate probes representing only the extremities of inserts. Where a restriction map is known, appropriate fragments can be isolated by preparative electrophoresis. Most replacement vectors have cleavage sites within their polylinkers, allowing clean separation of vector and insert DNAs. Inevitably, the ease of walking within any given chromosomal domain will be inversely related to the frequency of repetitive DNA it contains. Pre-annealing of a probe with a large excess of genomic DNA can effectively deplete a probe of repetitive sequences (27). Alternatively, probes can be pre-screened for repetitive sequences by Southern blotting (28). In the case of *Drosophila* it has been possible to take advantage of heterogeneity between different natural isolates (29). A walk initiated within one strain, upon encountering a repeated DNA sequence, can sometimes be continued in a second strain lacking a repeated DNA element at the chromosomal location in question.

The main drawback of the above methods is that at least some restriction mapping is necessary at each step of the walk. An alternative means of generating end-specific probes is to use a library in which promoters for phage-encoded RNA polymerases (e.g. T3, T7, and SP6 polymerases) flank the inserts. This allows the generation of RNA probes specific for either end of an insert. Though RNA probes are more susceptible to degradation than DNA probes, their preparation requires neither prior knowledge of restric-

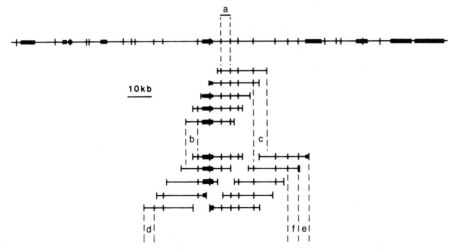

Figure 5. Chromosome walking. In this hypothetical case, a cloned fragment of single copy DNA (fragment **a**) is used to initiate a walk in both directions along the chromosome in which it normally resides. Vertical lines represent restriction sites. Each black box represents a member of a different family of repetitive DNA elements, other members of which lie elsewhere on the same or different chromosomes. Fragment **a** identifies a set of recombinant phage with overlapping inserts. Only the inserts are shown. Vertical lines within inserts correspond to restriction sites in the chromosome. Vertical lines at the ends of inserts correspond to restriction sites within the retained section of the polylinker. Fragments **b** and **c**, lying at the extremities of the newly cloned regions, are used to identify two further sets of overlapping inserts, and so on with fragments **d** and **f**. Fragment **e** cannot be used as it contains repetitive DNA sequences and will cause the walk to 'branch'. A very long section of repetitive DNA (at the extreme *right*) will be difficult to pass using a lambda library.

tion sites nor fragment isolation, thus significantly reducing the time and labour involved. Where a probe has a repetitive component it can be cleaned up in the same way as a DNA probe (30)

Even given improvements in vector design, long walks are still a formidable task. It would not be unusual to have to cover 1 Mb between a chromosomal entry point and a human disease gene, a distance that would require at least 100 steps in a lambda library (assuming an average insert size of 15 kb). An alternative to systematic walking through a large region is generation of a sub-library of lambda clones from a yeast artificial chromosome (YAC) that spans the region of interest. Overlap can be established quickly by hybridization of end-specific RNA probes representing individual clones to plaque lifts of a gridded array of the sub-library. This approach has been very successful for analysing a 400 kb region around the 5′ end of the human dystrophin gene (30). Gridded arrays, either of a sub-library or of an entire genomic library of an organism with a small genome, can be generated and replicated using robotic devices, or more cheaply using the plaque replicator described by MacKenzie *et al.* (31).

One disadvantage of an amplified library is the potential for misrepresentation due to phage to phage variation in growth rate. Although screening a primary library is clearly preferable, the phage would be used up quickly during repetitive screening. Taking several filter replicas of a primary library and probing each several times conserves a limited resource, but the need to recover viable phage at each step poses the problem that phage on conventional agar plates stored at 4 °C have a limited life. This can be overcome by using an agarose top layer containing glycerol and either storing the plates, or one set of plaque lifts, at −70 °C (32–35). Repeatedly screening filters lifted from the same plates also introduces a directional bias into a walk, since only newly identified clones need be picked at each step.

12. Bacteriophage lambda: relevant biology

The linear lambda genome is packaged in an icosahedral head, from which projects a tail. Adsorption to the host depends on interaction between the tail fibre, the product of the lambda gene *J*, and a cell surface receptor, the product of the *E. coli* gene *lamB*. It is the specificity of this interaction that determines host range. The receptors are a component of the *E. coli* maltose uptake pathway. Their production is stimulated by growth in medium containing maltose as a carbon source, and lacking glucose so as to avoid catabolite repression. Adsorption is temperature-independent and is facilitated by magnesium ions. Entry of the phage genome into a cell involves a component of the inner membrane. In contrast to adsorption it shows marked temperature-dependence, occurring rapidly at 37 °C and only slowly at low temperatures.

After entering a cell, the linear genome circularizes via its cohesive ends and covalent closure provides a substrate for replication (*Figure 7*). Wild-type lambda can follow either a lytic or a lysogenic pathway. In either case, transcription is initiated from two 'early' promoters, p_L and p_R (*Figure 6*), to provide functions essential for DNA replication, genetic recombination, establishment of lysogeny, and transcriptional activation of late genes. In lytic growth the transition from early to late transcription is achieved, virion proteins are made, and replicated genomes are packaged. Lysis of the cell releases approximately 100 infective particles. During lysogenic development most phage functions become blocked due to the action of repressor, the product of the *cI* gene. Repressor binds to operators associated with p_L and p_R, preventing RNA polymerase binding to either of these promoters. If repression occurs in time to prevent activation of late genes, lysis is avoided. Stable as opposed to abortive lysogeny also requires integration of the phage genome into the *E. coli* chromosome, where it is replicated passively as a prophage. Integration occurs by a site-specific recombination mechanism. None of the lambda derivatives described in this chapter has either a re-

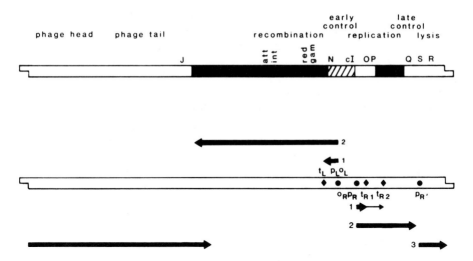

Figure 6. *Upper*: Genes that encode related functions are clustered in the lambda genome. *cI* together with genes in the black boxes are inessential for laboratory propagation of lambda, as is gene *N* in a phage lacking the termination site t_{R2} (see below). Deletion of these regions can thus provide space for exogenous DNA. Deletion of the repressor gene, *cI*, obliges vectors described here and their recombinant derivatives to grow lytically (*Figure 7*).

Lower: Transcription of the genome proceeds initially from two promoters, p_L and p_R. At the earliest stage of an infection (1), transcripts initiated at p_L and p_R terminate at t_L and t_{R1} respectively (some rightward transcripts escape termination at t_{R1} and reach t_{R2}). Only in the presence of the *N* gene product, an anti-terminator, does transcription proceed into adjacent genes (2). Consequent synthesis of the *Q* gene product allows 'late' transcription initiated at p_R' to continue through the lysis genes, past the *cos* site (of the circular genome: *Figure 7*), through the head and tail genes, and past the tail fibre gene *J* (3). Though *N* function is normally essential for late gene activation, deletion of t_{R2} allows *N*-independent 'leakage' of transcription through t_{R1} to provide sufficient *Q* function for late gene transcription to occur. Most lambda vectors, including those described here, make extra space via a deletion (*nin*5) that removes t_{R2}. The deletion makes *N* and the p_L promoter dispensable for phage growth. Even so all of the vectors described here, as well as their recombinant derivatives, are N^+. Because they do not have to rely on transcriptional leakage they grow relatively vigorously.

(●) Major promoters; (◆) major termination signals.

pressor gene or a site-specific integration system. All are thus obliged to grow lytically.

Bidirectional replication during the first 15 minutes of an infection yields monomeric circles (theta-mode replication; *Figure 7*). At later times the predominant replication products are linear concatemeric genomes, presumed to arise by a rolling circle mechanism. This change in mode of DNA replication coincides with activation of late genes, necessary for the production of virion proteins. Concatemeric DNA is the substrate for packaging, a

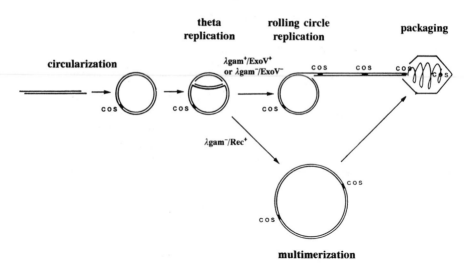

Figure 7. Lytic growth: Phage genomes can enter *E. coli* either as naked DNA (transfection) or by infection with a phage particle. Inside the cell the linear genome circularizes via its cohesive ends, and a covalently closed molecule is formed by the action of *E. coli* DNA ligase at the now double-stranded *cos* sites. During the first 15 minutes or so of an infection, bidirectional replication (theta replication) generates monomeric circular genomes. There is then a transition to rolling circle replication. If the phage lacks a functional *gam* gene the transition is blocked by the nucleolytic activity of *E. coli* Exo V (i.e. by RecBCD). *gam* is unnecessary for rolling circle replication in Exo V⁻ cells. The product of rolling circle replication is a linear DNA molecule, the usual substrate for assembly of DNA into phage particles. The packaging machinery recognizes two suitably spaced cos sites (40–52 kb apart) and makes staggered breaks, thereby recreating a linear genome bounded by cohesive ends. Although monomeric circles cannot be packaged, phage that lack a *gam* gene can be propagated in *recBC⁺* cells in the presence of a generalized recombination activity. Recombination between monomeric circles generates multimeric and thus packageable substrates.

process that involves the cutting of two appropriately spaced *cos* sites and incorporation of a linear genome into a phage head.

Transition to rolling circle replication is antagonized by exonuclease V, the product of the *E. coli recB, C*, and *D* genes (Exo V is is also known as the RecBCD nuclease). The nucleolytic activity of Exo V is, in turn, inhibited by the product of the lambda gene *gam*, expressed early during infection. The key role of *gam* is therefore obviated by the use of Exo V deficient hosts. All of the vectors described here are *gam⁺*, but most have their *gam* gene in the stuffer fragment (*Figure 8*).

gam⁻ phage can also grow in the presence of Exo V, if with less than wild-type yield, by a pathway involving recombination between the products of theta-mode replication (*Figure 7*). The resulting circular concatemer of phage genomes is a suitable substrate for DNA packaging. The lambda gene *red* provides a generalized recombination activity (*Figure 5*), but it is in the stuffer fragment of most of the vectors described here and is consequently absent from recombinants. Growth of $red^- gam^-$ phage in Exo V$^+$ bacteria thus depends on host recombination functions, the most efficient of which actually involves Exo V (Section 15.2 and *Figure 9*). It is paradoxical that Exo V both prevents $red^- gam^-$ phage from replicating by a rolling circle mechanism, and can also rescue them from inviability by promoting recombination.

Only phage genomes with one or more copies of an octanucleotide known as Chi are an efficient substrate for Exo V-mediated recombination, which also requires host RecA function (*Figure 9*). Though wild-type lambda has no Chi sites (Red is Chi-independent), EMBL vectors and their derivatives have been endowed with a Chi site in their right arm. Certain replacement vectors, such as Charon 4, Charon 4a, and Charon 30 (1, 2, 36), are both gam^- and Chi^- in the absence of the stuffer fragment. Recombinants that gain a Chi site are thus at considerable growth advantage over those that do not. Such vectors can be recommended only if the libraries will be recovered and amplified on Exo V$^-$ hosts. By contrast, Charon 34 and Charon 35 (37) (*Figure 8*) recombinants are gam^+, can therefore dispense with Chi, and can grow by rolling circle replication in the absence of any functional recombination system.

The recombination and nucleolytic activities of Exo V are separable. Unlike *recB* and *recC* mutations, *recD* mutations inactivate the nuclease while leaving the bacteria recombination proficient (38). Though a $recBC^-$ strain fosters large plaques, absence of the predominant genetic recombination pathway reduces bacterial vigour and consequently phage yield. Higher titres of $red^- gam^-$ phage, both on plates and in liquid culture, will be obtained in Rec$^+$ hosts, particularly $recD^-$ (Section 15.2).

13. Selection against reconstituted vector

gam^+ phage are unable to form plaques on *E. coli* lysogenic for phage P2 (2), the Spi$^+$ phenotype (sensitivity to *P2* interference). Though incompletely understood, this interaction can be used to advantage by including a functional *gam* gene in the stuffer fragment of an otherwise gam^- replacement vector (*Figure 8*). This, in principle, obviates the requirement for removal or inactivation of the stuffer fragment before ligation of vector and donor DNAs. Reconstituted vector genomes (gam^+) will be among the products of ligation, but can be discriminated against by propagating the library on an *E. coli* (P2) host. Only gam^-, and hence recombinant, phage should form plaques. The full Spi$^-$ phenotype also requires the phage to be red^- and Chi$^+$, conditions met by most replacement vectors. It should be borne in mind, however, that:

(a) Use of P2 lysogens as hosts limits both plaque size and the range of host genotypes. Neither $recBC^-$ (Rec$^-$, Exo V$^-$) nor $recD^-$ (Rec$^+$, Exo V$^-$) backgrounds can be used.

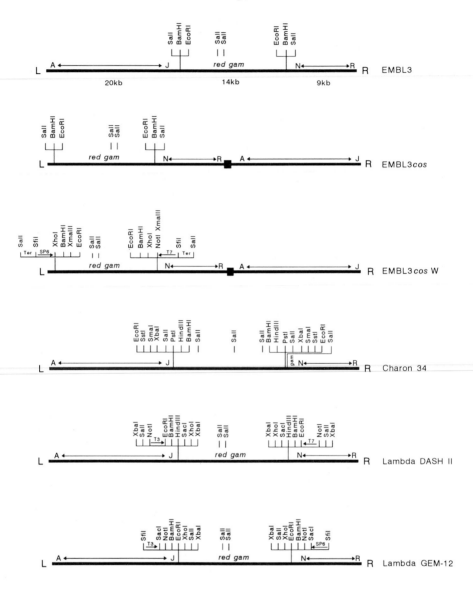

(b) Despite the apparent simplicity, libraries recovered by this strategy can have a high background of non-recombinants.

(c) Because of the above problem, it is likely that a method involving physical removal or inactivation of the stuffer fragment (Section 3.5) will be found preferable, particularly when working with limiting amounts of donor DNA (Section 6).

Figure 8. Replacement vectors discussed in the text.

EMBL3 (13) was derived in a number of steps from the vector 1059 (50). 1059 lacked polylinkers and had plasmid DNA sequences in its stuffer fragment. Reconstituted vector was thus a considerable problem when libraries were screened with plasmid probes. The stuffer of EMBL3 lacks plasmid sequences. Approximate sizes of the (L)eft and (R)ight vector arms and of the stuffer fragment are indicated. EMBL301 (51) (not shown) is a derivative of EMBL3 having more extensive polylinkers including sites for *Bgl*II, *Not*I/ *Xma*III, *Nae*I, *Sfi*I, and *Bgl*II/*Xho*I. EMBL4 (13) (not shown) is identical to EMBL3 except for the orientation of its polylinkers. A derivative of EMBL4, λRES (52), allows excision of a cloned fragment as a component of an autonomously replicating plasmid.

EMBL3*cos* (15) and EMBL3*cos*W (16) have all the essential phage genes in a long right arm, leaving the left cohesive end separated by only 200 bp from an insert (rather than by 20 kb as in the other vectors shown here). This arrangement has particular advantages for restriction mapping via partial cleavage of phage DNA labelled at its left end (Section 10). To make mapping even easier, a *Not*I site in the right-hand polylinker of EMBL3*cos*W allows labelled fragments to be cleanly separated from the right arm. SP6 and T7 promoters flanking the stuffer fragment of EMBL3*cos*W facilitate generation of RNA probes specific to one or other end of an insert. This has advantages for chromosome walking (Section 11). Terminators in the polylinkers of EMBL3*cos*W insulate vector genes from transcription initiated within cloned DNA. Black boxes are inactivated cos sites.

Charon 34 has sites in its polylinkers but not its arms for *Eco*RI, *Sst*I, *Xba*I, *Hin*dIII, and *Bam*HI (37). The presence of *gam* in the right arm rather than in the stuffer allows Charon 34 derivatives to grow on *recA⁻* hosts (Section 12). Charon 35 is identical to Charon 34 except for its stuffer fragment.

λDASH II (Stratagene) and λGEM-12 (Promega) have many of the qualities already discussed. Both have promoters for phage-specific RNA polymerases in their polylinkers. *Not*I sites flanking the T3 and T7 promoters of λDASH II allow an insert together with the two promoters to be isolated intact. *Sfi*I sites of λGEM-12 perform a similar function. Restriction maps can thus be generated by a partial cleavage strategy in which ends are labelled indirectly by probing a Southern blot with oligonucleotides based on the promoter sequences (λDASH II), or directly by ligation with labelled linkers complementary to either the left or right *Sfi*I terminus (λGEM-12) and then detecting the partial cleavage products by gel electrophoresis and autoradiography. λFIX II (not shown) is identical to λDASH II except for the composition of the polylinker. λGEM-11 (not shown) is identical to λGEM-12 except that it has an *Avr*II site in place of a *Not*I site.

These vectors are the sophisticated end-products of many years of genetic and bio-chemical manipulation of the lambda genome. For a complete description of their structure, the reader should refer to original sources. All the vectors shown are *cl⁻* (ΔKH54), and all carry the *nin*5 deletion (see legend to *Figure 6*), which removes the gene *rap* (see Section 9).

14. Choosing a vector

All of the replacement vectors in common use are capable of accommodating inserts of approximately 20 kb (*Figure 8*). All have *Bam*HI sites for cloning donor DNA fragments with ends generated by cleavage with *Sau*3A/*Mbo*I, and all have flanking sites that allow excision of an insert free of vector DNA sequences (on average only one in four *Bam*HI/*Sau*3A or *Bam*HI/*Mbo*I hybrid sites will be targets for *Bam*HI). Several vectors have promoters for phage-specific RNA polymerases, enabling the generation of RNA probes representing one or other extremity of an insert. This facility is extremely useful for chromosome walking (Section 11). λDASH II and λFIX II for example (available from Stratagene), have a T3 promoter in one polylinker and a T7 promoter in the other. λGEM-11 and 12 (available from Promega), and EMBL3*cos*W have SP6 and T7 promoters. The genome organization of EMBL*cos* vectors, in which the left cohesive end is within a few hundred base pairs of a polylinker, enables rapid restriction mapping (Section 10; see also ref. 16). All the vectors mentioned above have a stuffer fragment containing *red* and *gam*, thus allowing genetic selection against parental sequence phage. Recombinant derivatives of Charon 34 and 35 are *gam*$^+$, and may consequently be propagated in any Rec$^-$ host. Further discussion of the properties of specific vectors may be found in the legend to *Figure 8*.

15. Choosing a host

Examples of *E. coli* strains with characteristics discussed in this section are given in *Table 2*.

15.1 Restriction and modification

All strains liable to be used as hosts will be derivatives of *E. coli* K-12, which normally produces a restriction enzyme (*Eco*K) that will cleave DNA should it contain unmodified *Eco*K target sequences. *E. coli* K-12 protects its own genome by modifying (methylating) adenine residues within these specific seven bp sequences. Some recombinant genomes will not be so protected, even if the vector was grown in a modifying strain. Since the distribution of target sequences within a donor genome is unlikely to be random, this will cause misrepresentation of donor sequences in the library. The problem is

Table 2. *E. coli* hosts for propagation of recombinant phage

Strain[a]	Genotype
DL709	Δ*mcrA*, Δ (*mrr hsd mcrBC*), *sbcC*
NM767	Δ*mcrA*, Δ (*mrr hsd mcrBC*), *sbcC*, (P2*cox*3)
NM769	Δ*mcrA*, Δ (*mrr hsd mcrBC*), *sbcC*, *recD*1009
DL795	Δ*mcrA*, Δ (*mrr hsd mcrBC*), *sbcC*, *recA*::Cm^r
MN766	Δ*mcrA*, Δ (*mrr hsd mcrBC*), *sbcC*, *recA*::Cm^r/p*gam*
DL847	Δ*mcrA*, Δ (*mrr hsd mcrBC*), *sbcC*, *recD*1009, *recA*::Cm^r

[a] All of these strains (D. Leach and N. E. Murray, unpublished) are derived from DL709, an *mcrA* (e14°), Δ(*mrr hsd mcrBC*), *sbc*C201 derivative of *E. coli* K12SH28 (54). The only drug resistance is associated with the *recA* mutation, and is not part of a transposable element. Similar strains have been described elsewhere (43). The (*mcr hsd mcr*) deletion and *sbc*C201 are described in refs 42 and 48 respectively. NM538, recommended in ref. 1, is *mcrB*$^+$ (15); the *mcrA*$^-$*mcrB*$^-$ strain NM621 is now known to be *mrr*$^+$.

avoided by recovering the library in restriction deficient hosts. $hsdR^-$ strains are defective in restriction (rk$^-$ mk$^+$), while $hsdS^-$ strains are defective in both restriction and modification (rk$^-$ mk$^-$).

Other *E. coli* K-12 restriction systems attack DNA with methylated bases (39). These are of particular relevance to the generation of libraries of mammalian or plant DNA, and even some bacterial DNAs. Two activities are specific for DNA containing methylcytosine (McrA and McrBC), while a third (Mrr) attacks DNA with either methylcytosine or methyladenine residues (40, 41). In the context of a human DNA library, the use of $mcrB^+$ and $mcrA^+ mcrB^+$ hosts depressed the yield 10-fold and 30-fold respectively (15). The effect of Mcr was even more severe for plant libraries, and an additional small effect originally attributed to incomplete blockage of the McrBC system (42, 43) is now known to be a consequence of Mrr activity, but is designated McrF (41). The clustering of several restriction determinants at a single 'immigration control' locus allows the removal of *mrr* and *mcrBC* by a single deletion that also removes *hsd* (and thus rk and mk activities). A second deletion can be used to remove *mcrA* (41).

Unless the strains used to prepare *in vitro* packaging extracts lack the relevant restriction activities, restriction during packaging can be a problem. For *Eco*K, however, the consequence of five unmodified sites was only a two- to seven-fold reduction in packaging efficiency (44). The effects of the Mcr and Mrr systems have also been modest, except when the phage DNA was isolated from mammalian cells. In this case, the combined effects of *Eco*K, McrA, and McrF depressed the yield 1000-fold (45). Restriction during packaging is associated with the sonicated extract (56).

The rk phenotype is easy to check. Unmodified lambda form plaques three to four orders of magnitude more efficiently on rk$^-$ than on rk$^+$ hosts. Mcr activity can be detected by the restriction of glycosylation deficient T-even phages, when DNA containing hydroxymethylcytosine in place of cytosine becomes sensitive to Mcr restriction (39). T2 and T4 phages are restricted by both systems, T6 by only McrA.

Both the Mcr and Mrr activities are active on phage or plasmid DNA that has been modified by the methylase, M *Sss*I (41).

15.2 Genetic recombination

E. coli K-12 can sustain more or less efficient homologous recombination via a number of pathways (*Figure 9*); the RecBCD and RecF pathways both require the product of the *recA* gene, whereas the RecE pathway does not. The Red pathway is encoded by bacteriophage lambda. Only the RecBCD pathway is stimulated by Chi (Section 12).

Recombination properties are especially relevant to the choice of hosts for propagation and amplification of libraries. They are a major determinant of the general 'health' of bacterial strains, which in turn influences the phage yield (Section 12). For this reason, recombination proficient (Rec$^+$) hosts

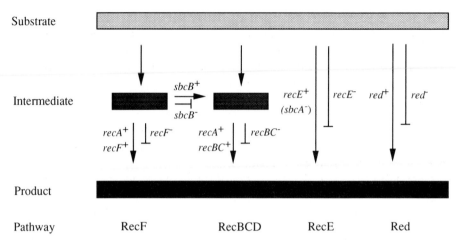

Figure 9. Recombination pathways of *E. coli* and lambda (adapted from ref. 53; see ref. 38 for a mini-review). The RecBCD pathway is the predominant (99%) recombination pathway of wild-type *E. coli* K-12. *recB⁻* and *recC⁻* single mutants, or *recB⁻ recC⁻* (*recBC⁻*) double mutants, are defective in this pathway. Presence of an *sbcB⁻* mutation in any of the above backgrounds appears to prevent shunting from the RecF to the RecBCD pathway, enabling efficient recombination (≈ 50% of wild-type) to be catalysed by the *recF* product. *recA⁻* strains, defective in both pathways, are the least able to catalyse recombination (10^{-3} to 10^{-6} of wild-type). Access to the RecE pathway is gained by an *sbcA* mutation. In terms of its biochemistry and genetics, the RecE pathway resembles, in many respects, the Red recombination pathway encoded by wild-type lambda. *recE* is carried by a defective lambdoid prophage present in the chromosome of most *E. coli* K-12 strains. Normally *recE* is repressed (as are the *red* genes of a lambda prophage). *sbcA* mutations, which map close to *recE*, in some way awaken it from repression. The *recBCD⁺*, *sbcB⁺*, and *recE⁺* products are respectively, exonucleases V, I, and VIII.

may be preferable, despite concern that homologous recombination between misaligned repetitive elements can cause duplications and deletions. Propagation in Rec⁻ hosts should circumvent the latter problem, but stability comes at the expense of more meagre yield of phage. This is due both to decreased burst size, and to the high proportion of dead cells in a Rec⁻ culture. Phage that adsorb to dead cells are lost. Furthermore, Rec⁻ strains tend to throw off more vigorous Rec⁺ revertants or pseudo-revertants. It is important, therefore, to be vigilant in checking the phenotype of Rec⁻ bacteria. Rec⁻ cells are impaired in their ability to repair UV induced damage to their DNA, a phenotype that is readily confirmed (see Section 2.97 in ref. 1).

The most stringent Rec⁻ condition is *recA⁻*. Although the *red⁻ gam⁺* derivatives of Charon 34 and Charon 35 can be propagated directly in *recA⁻* hosts, the *red⁻ gam⁻* derivatives of most other replacement vectors require recombination for growth unless the host also lacks the nucleolytic activity of Exo V. Unfortunately, *recA⁻ recBC⁻* and some *recA⁻ recD⁻* hosts are feeble

(Section 12). An alternative is a *recA⁻* host that carries a plasmid containing the *gam* gene of lambda (46). This plasmid is constructed such that expression of its *gam* gene is activated by the incoming phage. In this way, a *recA⁻ recBCD⁺* host can be used, the RecBCD nuclease will be inactivated by the product of the *gam* gene (Section 13.1), and phage replication will proceed normally. However, if this *recA⁻* host is used and the library is to be screened with a radioactive probe derived from a plasmid, care must be taken to avoid cross-hybridization between homologous plasmid sequences. The *recD⁻recA⁻* strain listed in *Table 2* appears to be as vigorous as the *recA⁻* strain and has given high titres of *red⁻gam⁻* phage.

Although one should be aware of recombinational instability, we would suggest that Rec⁻ hosts are used only as a last resort. To grow *red⁻ gam⁻* recombinants under conditions of recombination proficiency, it is often preferable to use a background that is also permissive for phage DNA replication by the rolling circle mode. *recD⁻* hosts, deficient in the nucleolytic but not the recombinational activity of the RecBCD (Section 12), are quite vigorous, as are both *recBC⁻sbcA⁻* hosts (proficient in the RecE pathway) and *recBC⁻ sbcB⁻* hosts (proficient in the RecF pathway) (*Figure 9*).

Finally it should be noted that long palindromic sequences are unstable in *E. coli*. Use of *recBC⁻ sbcB⁻* hosts had been recommended as a means of circumventing this problem (47). Such strains all carried the additional mutation *sbcC⁻*, however, which now appears to be the mutation of relevance (48).

Taking all of the above observations to their logical conclusion, the following genetic backgrounds are likely to be most suitable for the propagation of *red⁻ gam⁻* phage: either *sbcC⁻* or *recD⁻ sbcC⁻* for propagation under conditions of recombination proficiency; and either *recA⁻ recD⁻ sbcC⁻* or *recA⁻ sbcC⁻/pgam* for propagation under conditions of recombination deficiency. Hosts for the initial recovery of packaged genomes as phage should also be deficient in the restriction activities discussed in Section 15.1. Recommended strains are NM772 (Rec⁺) and DL847 (Rec⁻) (*Table 2*). Even so, no host guarantees the recovery of a specific class of recombinant.

16. Why lambda?

The major advantage of cosmid, phage P1, and YAC vectors is their greater capacity. Not only are less recombinants required for a representative library, but less steps are required for a chromosome walk, and there is a greater chance that a complete gene will be present within a single recombinant. Disadvantages are that libraries are generally more difficult to construct and to screen, and that large inserts are difficult to analyse. Lambda, with its smaller insert size, imposes a less stringent requirement on the quality of the donor DNA. Very efficient recovery of recombinant molecules is possible by *in vitro* packaging. Moreover, lambda libraries remain the easiest to screen by hybridization. Lysed cells release large amounts of unpackaged DNA along with phage.

Although recent improvements to cosmid vectors allow libraries to be constructed from smaller amounts of donor DNA (49), the same cannot be said of P1 or YAC vectors. Complications with the analysis of YACs arise from the size of the inserts and the difficulty of isolating YACs from a yeast background. YACs thus benefit from subcloning into phage vectors (21, 30).

In summary, the utility of lambda resides in the ease of library construction, screening, and analysis. It still has an important part to play in many cloning strategies.

References

1. Sambrook, J., Fritsch, E. F., and Maniatis, T. (ed.) (1989). *Molecular cloning: a laboratory manual* (2nd edn). Cold Spring Harbor Press, NY.
2. Hendrix, R. W., Roberts, J. W., Stahl, F., and Weisberg, R. A. (ed.) (1983). *Lambda II*. Cold Spring Harbor Press, NY.
3. Bothwell, A., Yancopoulos, G. D., and Alt, F. W. (1990). *Methods for the cloning and analysis of eukaryotic genes*. Jones and Bartlett, Boston, USA.
4. Riley, J. H., Ogilvie, D., and Anand, R. (1992). In *Techniques for the analysis of complex genomes* (ed. R. Anand). Academic Press, New York.
5. Sternberg, N. (1992). *Trends Genet.*, **8**, 11.
6. Sanger, F., Coulson, A. R., Hong, G.-F., Hill, D. F., and Petersen, G. B. (1982). *J. Mol. Biol.*, **162**, 301.
7. Tilzer, L., Thomas, S., and Moreno, R. F. (1989). *Anal. Biochem.*, **183**, 13.
8. Sealey, P. G. and Southern, E. (1982). In *Gel electrophoresis of nucleic acids: a practical approach* (ed. D. Rickwood and B. D. Hames), pp. 39–76. IRL Press, Oxford.
9. Maniatis, T., Hardison, R. C., Lacy, E., Lauer, J., O'Connell, C., Quon, D., Sim, D. K., and Efstratiadis, A. (1978). *Cell*, **15**, 687.
10. McClelland, M. and Nelson, M. (1991). *Nucleic Acids Res.*, **19**, 2045.
11. Coulson, A. and Sulston, J. (1988). In *Genome analysis: a practical approach* (ed. K. E. Davies), pp. IRL Press, Oxford.
12. Bellane-Chantelot, C., Barillot, E., Lacroix, B., le Paslier, D., and Cohen, D. (1991). *Nucleic Acids Res.*, **19**, 505.
13. Frischauf, A.-M., Lehrach, H., Poustka, A., and Murray, N. (1983). *J. Mol. Biol.*, **170**, 827.
14. Zabarovsky, E. R. and Allikmets, R. L. (1986). *Gene*, **42**, 119.
15. Whittaker, P. A., Campbell, A. J. B., Southern, E. M., and Murray, N. E. (1988). *Nucleic Acids Res.*, **16**, 6725.
16. Whittaker, P. A. (1991). *Clin. Biotech.*, **3**, 67.
17. Fuchs, R. and Blakesley, R. (1983). In *Methods in enzymology* (ed. R. Wu, L. Grossman, and K. Moldave), Vol. 100, pp. 3–38. Academic Press, New York.
18. Clarke, L. and Carbon, J. (1976). *Cell*, **9**, 91.
19. Zilsel, J., Ma, P. H., and Beatty, J. T. (1992). *Gene*, **120**, 89.
20. Hoheisel, J. D., Nizetic, D., and Lehrach, H. (1989). *Nucleic Acids Res.*, **17**, 4571.

21. Whittaker, P. A., Mathrubutham, M., and Wood, L. (1993). *Trends Genet.*, **9**, 195.
22. Seed, B. (1983). *Nucleic Acids Res.*, **11**, 2427.
23. Kurnit, D. M. and Seed, B. (1990). *Proc. Natl. Acad. Sci. USA*, **87**, 3166.
24. Lutz, C. T., Hollifield, W. C., Seed, B., Davie, J. M., and Huang, H. V. (1987). *Proc. Natl. Acad. Sci. USA*, **84**, 4379.
25. Rackwitz, H.-R., Zehetner, G., Frischauf, A.-M., and Lehrach, H. (1984). *Gene*, **30**, 195.
26. Rommens, J. M., Ianuzzi, M. C., Kerem, B. S., Drumm, M. L., Melmer, G., Dean, M., Rozmahel, R., Cole, J. L., Kennedy, D., Hidaka, N., Zsiga, M., Buchwald, M., Riordan, J. R., Tsui, L. C., and Collins, F. S. (1989). *Science*, **245**, 1059.
27. Sealey, P. G., Whittaker, P. A., and Southern, E. M. (1985). *Nucleic Acids Res.*, **13**, 1905.
28. Howell, M. D., Resner, J., Austin, R. K., and Kagnoff, M. F. (1987). *Gene*, **55**, 41.
29. Bender, W., Spierer, P., and Hogness, D. S. (1983). *J. Mol. Biol.*, **168**, 17.
30. Whittaker, P. A., Wood, L., Mathrubutham, M., and Anand, R. (1993). *Genomics*, **15**, 453.
31. MacKenzie, C., Stewart, B., and Kaiser, K. (1989). *Technique*, **1**, 49.
32. Klinman, D. M. and Cohen, D. I. (1987). *Anal. Biochem.*, **161**, 85.
33. Whittaker, P. A. and Lavender, F. L. (1989). *Nucleic Acids Res.*, **17**, 4406.
34. Whittaker, P. A. (1993). In *Methods in enzymology* (ed. R. Wu), Vol. 218, pp. 366–627. Academic Press, New York.
35. Requena, J. M., Sot, M., and Alonso, C. (1993). *Trends Genet.*, **9**, 4.
36. Murray, N. E. (1991). In *Methods in enzymology* (ed. J. Miller), Vol. 204, pp. 280–304. Academic Press, New York.
37. Loenen, W. A. M. and Blattner, F. R. (1983). *Gene*, **26**, 171.
38. Smith, G. (1989). *Cell*, **58**, 807.
39. Raleigh, E. A., Murray, N. E., Revel, H., Blumenthal, R. M., Westaway, D. A., Reith, A. D., Rigby, P. W. J., Elhaai, J., and Hanahan, D. (1988). *Nucleic Acids Res.*, **16**, 1563.
40. Waite-Rees, P. A., Keating, C. J., Moran, L. S., Slatko, L. J., and Benner, J. S. (1991). *J. Bacteriol.*, **173**, 5207.
41. Kelleher, J. E. and Raleigh, E. A. (1991). *J. Bacteriol.*, **173**, 5220.
42. Woodcock, D. M., Crowther, P. J., Doherty, J., Jefferson, S., DeCruz, E., Noyer-Weidner, M., Smith, S. S., Michael, M. Z., and Graham, M. W. (1989). *Nucleic Acids Res.*, **17**, 3469.
43. Doherty, J. P., Lindeman, R., Trent, R. J., Graham, M. W., and Woodcock, D. M. (1993). *Gene*, **124**, 29.
44. Rosenberg, S. M. (1985). *Gene*, **39**, 313.
45. Gossen, J. A. and Vijg, J. (1988). *Nucleic Acids Res.*, **16**, 9343.
46. Crouse, G. (1985). *Gene*, **40**, 151.
47. Leach, D. R. F. and Stahl, F. W. (1983). *Nature*, **305**, 448.
48. Chalker, A. F., Leach, D. R. F., and Lloyd, R. G. (1988). *Gene*, **71**, 201.
49. Evans, G. A., Lewis, K., and Rothenberg, B. E. (1989). *Gene*, **79**, 9.
50. Karn, J., Brenner, S., Barnett, L., and Cesareni, G. (1980). *Proc. Natl. Acad. Sci. USA*, **77**, 5172.

51. Lathe, R., Villotte, J. L., and Clark, A. J. (1987). *Gene*, **57**, 193.
52. Altenbuchner, J. (1993). *Gene*, **123**, 63.
53. Horii, Z. and Clark, A. J. (1973). *J. Mol. Biol.*, **80**, 327.
54. Fangman, W. and Novick, A. (1966). *J. Bacteriol.*, **91**, 2390.
55. Manfioletti, G. and Schneider, C. (1988). *Nucleic Acids Res.*, **16**, 2873.
56. Gunther, E. J., Murray, N. E., and Glazer, P. M. (1993). *Nucleic Acids Res.*, **21**, 3903.
57. Birren, B. and Lai, E. (1993). *Pulsed field gel electrophoresis: a practical guide*. Academic Press, New York.

Procedures for cDNA cloning

CHRISTINE J. WATSON and JEROME DEMMER

1. Introduction and background

The first mammalian cDNA was cloned almost two decades ago (1) and the construction and screening of cDNA libraries has since become one of the fundamental procedures of molecular biology. A considerable number of advances have been made during the past 20 years, including the availability of highly purified, cloned, and modified enzymes which has improved the efficiency of cDNA synthesis and cloning. The development of high efficiency cloning systems has allowed the construction of a representative cDNA library to become a straightforward procedure which can be successfully carried out in any molecular biology laboratory. More recently, the polymerase chain reaction (PCR) has revolutionized cDNA cloning and in some circumstances it will not be necessary to construct a library for the isolation of a particular cDNA clone.

In this chapter, we focus first on the construction of representative cDNA libraries in a suitable lambda or plasmid vector, and secondly describe how PCR can be used to clone a partial cDNA. Protocols for the construction of a standard cDNA library will be presented and where appropriate alternative procedures and reagents that are available will be discussed. A protocol for PCR cloning of a particular cDNA is presented and problems that may be encountered are highlighted. A number of ingenious procedures have been devised for cloning individual cDNAs and it is beyond the scope of this chapter to discuss all of these in detail.

It is possible to purchase individual reagents for the synthesis and cloning of cDNA. Alternatively, complete commercial kits for making either cDNA libraries or cloning by PCR are available. The majority of these kits are very good and the extent to which they are used depends very much on personal circumstances. Important considerations are the individual's technical expertise and experience, the number and size of libraries that are required, the length of mRNA to be cloned, the importance of full-length clones, reagents and equipment that are already available, and of course the cost. If after reading this chapter the prospect of making a cDNA library is too daunting you can buy a pre-made library or have one custom-made by a commercial company (e.g. Stratagene, Clontech, Invitrogen). Throughout this chapter commercial kit alternatives to particular steps in the cloning procedure will be mentioned.

2. Considerations for devising a cloning strategy

The first choice to be made is between constructing a cDNA library or cloning by reverse transcription coupled PCR (RT-PCR). If some sequence information is available then it may be possible to clone the cDNA by PCR. This is discussed in Section 5. It is worth making an attempt to clone the desired cDNA using PCR if at all possible since this approach is much quicker and simpler than building and screening a large cDNA library. If this fails or if a longer cDNA clone is required, then a conventional library can be made. In all other situations it will be necessary to construct a library. The factors that have to be considered before deciding upon a cloning strategy are discussed below.

2.1 Source and relative abundance of the particular mRNA

The source of mRNA may be obvious. The required transcript may, for example, be expressed in a tissue-specific manner. RNA from tissues or cells containing the highest level of the specific mRNA(s) of interest should be used for cDNA synthesis provided the RNA can be extracted in an undegraded form. Some tissues, for example the pancreas, contain high levels of ribonucleases (RNases) which can degrade RNA rapidly. It may be possible to increase the abundance of rare mRNAs by treating a tissue or cells with an agent to induce expression of the gene of interest, or by using tumour tissue in which the particular mRNA is overexpressed. The mRNA for preproenkephalin for example, is expressed at high levels in pheochromocytomas, tumours of the adrenal gland (2).

The abundance of the mRNA will determine the size of cDNA library which must be made to ensure that an appropriate clone can be isolated. Low abundance mRNAs are defined as those which are present at less than about 20 copies per cell. This class of RNA usually constitutes about 90% of the expressed sequences and 30% of the mRNA. The number of clones required to give a 99% probability that a particular rare mRNA will be present in a cDNA library can be calculated according to the Clarke and Carbon equation (3):

$$N = \ln (1 - P) / \ln (1 - 1/n)$$

where N = the number of clones required, P = the desired probability (99%), and $1/n$ = the fractional proportion of the rare mRNA in the total pool of mRNA. These numbers have been calculated for a number of cell types. The example below is for a rare mRNA (14 copies/cell) in a fibroblast cell line (4):

$$P = 0.99$$
$$1/n = 1/37\,000$$
$$N = 170\,000.$$

In practice, a larger number of recombinant clones will have to be screened to ensure isolation of a rare cDNA. A number of factors will influence the representation of a particular cDNA clone in a library. These include mRNA instability, the tendency of some sequences to be preferentially cloned, and the presence of multiple cell types in a particular tissue some of which will not contain the RNA in question. Moreover, some mRNAs will be especially rare.

Constructing large, representative cDNA libraries is easier if the source of starting material is not limiting. Where this is not the case, for example with early mouse embryos or rare tumour samples, amplification of the first strand cDNA by PCR may be necessary (see Section 5). It should be borne in mind that screening large numbers of clones (in excess of 10^6) is a laborious process and requires large amounts of materials. Depending on the screening procedure to be used this may prove difficult and expensive.

2.2 Length of mRNA

The length of the specific mRNA which is to be cloned will influence the choice of primer and vector, and will also determine the need for size selection. The majority of mRNA species are between 1.5 and 2.0 kb in length. For these 'standard' mRNAs the primer of choice is oligo(dT) which is 12–18 nucleotides in length. The oligo(dT) primer will anneal to the poly(A) tract which is found at the 3' terminus of most eukaryotic mRNAs (histones are a notable exception). There are, however, a number of mRNAs which are extremely long and it would be difficult to synthesize complete cDNA copies from these long RNAs in one piece. This problem can be overcome by the use of mixed hexanucleotide primers to allow random priming of cDNA along the mRNA strand which can then be cloned as a number of segments. A better option may be to use a specific primer for the mRNA of interest but this requires some knowledge of the sequence. This can be a useful approach when the RNA of interest is a member of a superfamily of genes which have short regions of high sequence conservation. It must be noted, however, that cDNA synthesis using such gene-specific primers is not always successful.

If the size of the mRNA is larger than 2.5 kb, it may be worthwhile to enrich for either the starting mRNA or preferably the double-stranded cDNA synthesized from it. This will reduce both the size of library that has to be made and the number of clones that have to be screened. Fractionating double-stranded cDNA is preferred because mRNA is highly susceptible to degradation by ribonucleases which would reduce the chance of cloning a full-length corresponding cDNA. Moreover, because the mRNA will have to be purified from agarose gels, any remaining agarose can inhibit the activity of enzymes and thereby reduce the efficiency of cDNA synthesis.

It should be remembered that there is an upper limit to the size of cDNA insert which can be cloned into lambda cloning vectors which varies between 7 kb and 12 kb. If larger cDNA inserts are required a plasmid or cosmid vector should be used. In practice, this size limit is not a problem as it is difficult to obtain large (> 7 kb) double-stranded cDNA molecules.

2.3 Screening procedure

The screening procedure to be utilized will determine whether lambda or plasmid vectors are used and which particular cloning vector should be selected (see Section 3.1 and *Table 1*). If nucleic acid probes are to be used, a vector such as λgt10 is preferred because of the large clear plaques which can be obtained. Screening plaques is generally a lot easier and cleaner than colonies. For expression cloning, which depends on the production of a fusion protein in the plaques *in situ*, then the λgt11 vector or one of its derivatives should be used. If eukaryotic expression is required to identify the clones of interest, then it will be necessary to use a plasmid vector containing an appropriate promoter (e.g. CMV). If the screening procedure requires the expression of antisense or sense copies of the mRNA then directional cloning will be essential. Directional cloning is also necessary for subtraction cloning (see Section 4.3). It is worth noting that when the cDNA is inserted into an expression vector in the sense orientation (i.e. the correct orientation) to produce copies of the mRNA, this will increase the likelihood of obtaining the correct fusion protein by twofold. This may not seem much but is relevant when considering the difference between screening one million and two million plaques!

3. Standard library construction

The choice of cloning strategy should be based on the considerations described above. For most purposes, a standard library in a lambda vector is all that will be required. This section will present detailed protocols for making such a standard cDNA library, and is an updated and revised version of our previous cloning protocol (5). *Figure 1* is a flow diagram of the steps involved. At each step there are a number of options. The use of state-of-the-art cloning vectors and modified enzymes can offer considerable advantages over the original procedure and it is certainly worth considering their use. We will discuss each procedure and the alternatives, including cloning into a plasmid vector, in detail below.

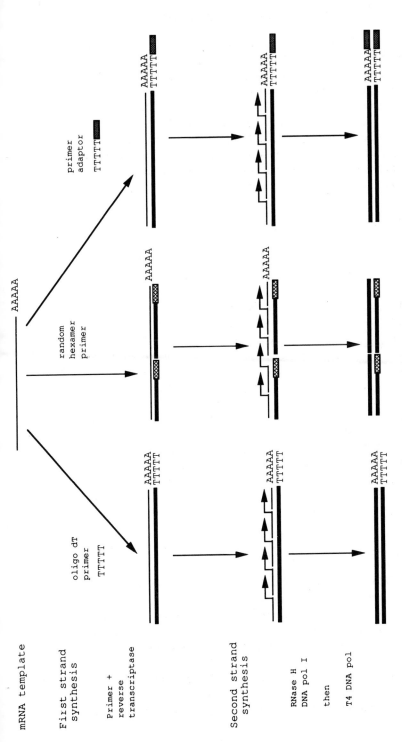

Figure 1. Flow diagram of a standard cDNA cloning procedure. The first cDNA strand is synthesized using a short primer. There are three different types of primers that can be utilized: an oligo(dT) primer will anneal with the poly(A) tail at the 3' end of the mRNA (it should be noted that the primer may also anneal to an internal stretch of A residues); random primers will anneal at many positions along the length of the mRNA template resulting in shorter cDNAs; a primer adaptor will provide a restriction site at one end of the cDNA. The second strand is synthesized using RNase H generated mRNA primers and DNA polymerase I. The double-stranded cDNA is then repaired and made blunt-ended with T4 DNA polymerase. The cDNA is now ready for the addition of linkers or adaptors, size selection, and ligation to the vector of choice.

3.1 Preparation of vector DNA

The first choice to be made is the cloning vector. Lambda vectors offer the advantages of high efficiency cloning, easy screening and storage and, until recently, superceded the use of plasmid vectors. This situation has changed with the availability of improved plasmid vectors which, when coupled with the technique of electroporation into competent *E. coli* cells, allow similar cDNA cloning efficiencies to *in vitro* packaging with phage lambda. The advantages of electroporation of plasmids are the elimination of subcloning and the relative ease of plasmid manipulation. One drawback is that a suitable electroporation (pulser) apparatus is essential for transforming bacteria with the efficiency necessary for making a cDNA library. This equipment may already be available since electroporation is frequently used to transfect cell culture and embryonic stem (ES) cells. The pulser itself is expensive to buy and purchasing lambda packaging extracts will be a much cheaper alternative.

We have frequently used the phage vector λgt10 for libraries to be screened with nucleic acid probes. Unless the library is required for just one specific purpose it is better to use an expression vector such as λgt11. Expression libraries can be screened with nucleic acid probes, antibody probes and, for the cloning of some transcription factors, by affinity screening with cognate oligonucleotide recognition sites. If the screening procedure requires expression in yeast or mammalian cells then it is preferable to clone the cDNA directly into a suitable expression plasmid. Alternatively, a phagemid vector with an *in vivo* excision capability can be used.

A variety of suitable vectors in addition to λgt10 and λgt11 are now commercially available. These include λZiplox (Gibco-BRL, Cat. No. 530–5395SA), λZAP II (Stratagene, Cat. No. 236201), and the directional vectors λMax 1, λYES (Clontech, Cat. No. 6192–1 and 6163–1), UniZAP XR, ZAP Express (Stratagene, Cat No. 237211 and 239201), as well as LambdaGEM-2 and -4 (Promega, Cat. No. T3441 and T3451). A compilation and comparison of cloning vehicles is presented in *Table 1*. This list is not exhaustive. The basic λgt10 and λgt11 vectors are perfectly adequate for constructing representative libraries and can be purchased from a number of companies. Their commercial derivatives offer a number of advantages with the only real disadvantage being the cost. Subcloning of cDNA inserts is easy with the vectors that permit *in vivo* excision (e.g. ZAP, Express, Ziplox) and works well in our hands. Directional cloning is necessary for expression cloning. The LambdaGEM vectors contain the T7 and SP6 promoters allowing RNA to be transcribed from the clones which could be useful in differential screening.

The choice of vector really depends on the requirements for screening (as discussed above), the cost, and what is already available. If full-length clones are essential, then it is worth considering the plasmid-based vector in the cDNA ClonStruct™ Kit from United States Biochemicals (discussed in

Table 1. Comparison of lambda insertion vectors and plasmid vectors

Lambda vectors[a]	Cloning capacity	*In vivo* excision	Directional cloning	β-gal fusion protein	Antibiotic resistance
λgt10	7.6 kb	No	No	No	None
λgt11	7.2 kb	No	No	Yes	None
λgt22A	8.2 kb	No	Yes	Yes	None
λZiplox	7.0 kb	Yes	Yes	Yes	AMP
λZAP II	10.0 kb	Yes	Yes	Yes	AMP
UniZAP XR	10.0 kb	Yes	Yes	Yes	AMP
λExpress	12.0 kb	Yes	Yes	Yes	KAN
λMax1	8.0 kb	Yes	Yes	No	AMP
λYES	8.0 kb	Yes	Yes	No	AMP
λDR2	7.0 kb	Yes	Yes	No	AMP
λBlueMid	7.0 kb	No	Yes	No	AMP
λGEM-2	7.1 kb	No	Yes	No	None
λGEM-4	4.7 kb	No	Yes	No	None

Plasmid/ phagemid vectors[a]	Cloning sites	Eukaryotic expression	Directional cloning	β-gal fusion protein	Antibiotic resistance
pBK-CMV	17	Yes	Yes	Yes	KAN, G418
pBK-RSV	17	Yes	Yes	Yes	KAN, G418
pBluescript II	21	No	Yes	Yes	AMP
pDR2	3	Yes	Yes	No	AMP, HYG
pcDNA 1	8	Yes	Yes	No	AMP (by *sup*F)
pCDM8	8	Yes	Yes	No	AMP (by *sup*F)
pYEUra3	5	Yes	Yes	Yes/No	AMP

[a] The suppliers of these vectors are as follows:
Stratagene: λZap II, UniZap XR, λExpress, pBK-CMV, pBK-RSV, pBluescript II.
Promega: λGEM-2, λGEM-4.
Clontech: λMax 1, λYES, λDR2, λBlueMid, pDR2, pYEUra3, pCDM8, pcDNA 1.
Gibco-BRL: λZiplox, λgt22A.

Section 3.4.1). It should be noted that the phagemid pBK-CMV, which can be used directly for cDNA cloning, is excised *in vivo* from the λExpress vector, as is pDR2 from the λDR2 vector, and pYEUra3 from lambda Max 1. The following are some general guide-lines for vector selection based on sceening requirements:

- use λgt10 for DNA probes
- use λgt11 for antibodies and recognition site oligonucleotides
- use plasmids for expression or complementation in yeast
- use plasmids for expression in tissue culture cells
- if an electroporator is not available use λExpress or λDR2
- use the appropriate state-of-the-art vector if affordable—this will increase the utility and flexibility of the library.

Once a suitable vector has been chosen, good quality vector DNA should be prepared. Standard CsCl banding of plasmid DNA (6) is recommended since it is important that it is essentially free of nicks. A detailed protocol for preparing lambda DNA can be found in Chapter 2 and in ref. (7). Dephosphorylation of the vector, to increase the ratio of recombinants to non-recombinants, may or may not be necessary depending on the vector type. With λgt10, non-recombinants can easily be selected against by using the *E. coli* C600Hfl150 strain, in which non-recombinants will lysogenize the bacteria and therefore do not form plaques. The tac-cI selection system is utilized in λDR2 to give almost 100% recombinant clones (8). For λgt11 and related vectors, dephosphorylation is recommended. Directional cloning, using different restriction sites at either end, is another way in which to reduce vector self-ligation.

3.2 Preparation of mRNA template

High quality mRNA is essential for the construction of libraries containing full-length inserts. It is worth ensuring that the mRNA is not degraded and is free of contaminants, such as phenol and SDS, that may inhibit the synthesis of cDNA.

3.2.1 Isolation of total RNA

There are a variety of methods for the isolation of total RNA from cells or tissues. Commercial kits are available from many companies including Pharmacia, Stratagene, and Gibco-BRL. *Protocol 1* gives a procedure for isolating total RNA from most tissues. The following technical tips should be followed to reduce the risk of RNase contamination:

- all glassware should be baked at 180 °C for at least 4 h
- all plasticware should be new and sterile; dispose after use
- deionized distilled water (ddH$_2$O) should be treated with diethylpyrocarbonate (DEPC) before use. Add DEPC to 0.1%, leave for at least 12 h, and then autoclave. Caution: DEPC is carcinogenic, toxic, and may explode upon opening of the bottle.
- gloves must be worn at all times.

Protocol 1. Isolation of RNA from cells or tissues

Reagents

- Appropriate tissue or cells
- Phosphate-buffered saline (PBS) (use PBS tablets (Oxoid) dissolved in ddH$_2$O and then treat with DEPC as described)
- RNazol B (Cinna/Biotecx Laboratories Inc., Houston, Texas, USA)
- Liquid nitrogen
- 100% ethanol
- 75% ethanol (in which the ddH$_2$O should have been DEPC treated)

Method

1. Flash freeze tissue samples in liquid nitrogen as quickly as possible following removal from the animal. For tissue culture cells, remove the medium, wash the cells with ice-cold PBS, scrape into a centrifuge tube, and collect by centrifugation at 2500 r.p.m. for 10 min. Wash once more with ice-cold PBS, centrifuge as before, then flash freeze.

2. Grind the tissue to a fine powder under liquid nitrogen using a mortar and pestle. Extract the RNA with RNazol using 2 ml/100 mg of tissue, or 200 μl/10^6 cells.

3. Homogenize using a polytron or an Ultra-turrax T25 homogenizer until completely solubilized.

4. Split the homogenates into two 1 ml aliquots in 1.5 ml microcentrifuge tubes, and add 100 μl of chloroform to each tube.

5. Vortex for 15 sec.

6. Put on ice for 15 min.

7. Microcentrifuge for 15 min at 4 °C and transfer the upper aqueous phase to a fresh tube. Take 500 μl at most from each tube, taking care to keep the pipette tip away from the interface which contains RNase and DNA. Precipitate with an equal volume of isopropanol for 45 min at -20 °C.

8. Microcentrifuge for 15 min at 4 °C and pour off the supernatant. Drain the tube on tissues. Watch that the pellet does not slip out of the tube!

9. Wash the pellet twice with 1 ml of 75% ethanol by vortexing and centrifuging for 8 min at room temperature. Pour off the supernatant and drain as above.

10. Dry under vacuum for 10 min, but do not let the RNA dry out completely as this will decrease its solubility.

11. Dissolve the RNA in 200 μl of DEPC treated ddH$_2$O. This may require heating to 60 °C and vortexing.

12. Read the OD 260/280 ratio of a diluted aliquot. This should be greater than 1.9. The sample used for this purpose should be discarded.

3.2.2 Selection of polyadenylated RNA (mRNA)

Over 95% of the total RNA isolated from a cell or tissue will be ribosomal RNA. Although it is possible to make libraries from total RNA, we would recommend where possible the use of polyadenylated RNA otherwise self-primed rRNA clones will constitute part of the library. The conventional method for the selection of polyadenylated RNA is by chromatography on oligo(dT) cellulose (9). This is a laborious procedure and an alternative is to use spin columns which are available as kits from a number of companies (e.g.

Pharmacia mRNA purification kit, Cat. No. 27–9258–01). Spin columns should be used according to the manufacturers' instructions. Another alternative is to select mRNA using oligo(dT) coupled to magnetic beads. These magnetic beads are available from Dynal or Promega and allow a one-step purification of mRNA from tissue.

3.2.3 Assessing the integrity of the mRNA

Before starting cDNA synthesis, it is vital to check the integrity of the mRNA which is to be used as the template. If the RNA appears to be even slightly degraded, a fresh preparation should be made. If the mRNA is not going to be used within a short time, it is best stored in 75% ethanol at $-70\,^\circ$C. There are a number of ways in which the integrity of mRNA can be assessed. The simplest of these is by gel electrophoresis followed by northern blot analysis for which a protocol may be found in ref. 6. Briefly, a proportion of the mRNA (usually 2 µg is sufficient) is electrophoresed on a denaturing agarose gel with a track of RNA size markers (e.g. Pharmacia, Boehringer) and the gel stained with ethidium bromide (Caution: ethidium bromide is a mutagen). A smear should be seen from about 8 kb down to about 400 bp, with the greatest intensity between 1.5 kb and 2.0 kb, although this will vary for different tissues. The gel can then be blotted and hybridized with a suitable probe representing a transcript known to be fairly abundant in the mRNA sample under test. Usually, β-actin is suitable for this purpose. Hybridization may also be carried out with the probe which will be used to screen the library and this will give an indication of the level of the mRNA of interest, whether it is degraded, and will determine how useful the probe will be in screening the cDNA library. An alternative is to translate the mRNA *in vitro* using either a reticulocyte lysate or wheat germ translation system (6).

3.3 Synthesis of cDNA

Synthesis and cloning of cDNA involves a number of steps and there are a variety of alternative procedures for each of these. The protocol presented here is optimal in our hands, and is detailed below, along with suitable controls for each stage to assess the quality and quantity of the product. The protocol is a 'one tube' method where both first and second strand reactions are carried out in a single tube.

3.3.1 First strand cDNA synthesis

The first step in the synthesis of cDNA is copying the mRNA template into complementary single-stranded cDNA. The enzyme used for this purpose is reverse transcriptase (RT), an RNA-dependent DNA polymerase, which extends a DNA primer annealed to an mRNA template. The RT purified from an avian retrovirus (AMV) was used for the original cloning experiments, but this enzyme has largely been superceded by Moloney murine leukaemia virus RT (M-MLV RT) because the latter enzyme has considerably lower levels of endogenous RNase H activity which can cause degradation of the template before full-length cDNA has been synthesized. First strand reaction conditions are given in *Protocol 2*. It is essential to add the RNase inhibitor RNasin (Boehringer) to improve the length of first strand transcripts when using AMV RT, but this is optional if M-MLV RT is used. A number of options are available at this stage:

(a) Primers other than oligo(dT) can be used. For directional cloning a synthetic oligonucleotide primer composed of a restriction enzyme site linked (5′) to about 12–18 T residues should be used (e.g. an *Xba* I oligo (dT) primer adaptor GTCGACTCTAGA(T)$_{15}$).

(b) Random hexamer primers increase the likelihood of obtaining the 5′ ends of the mRNA and can be used alone or mixed with oligo(dT).

(c) A different RT can be used. Superscript RT plus (Life Technologies) is a modified M-MLV RT which lacks RNase H activity thereby increasing the length of first strand cDNA. This enzyme also has increased thermostability (up to 50 °C). Carrying out a synthesis reaction at this temperature should reduce secondary structure in the RNA template which may cause the RT to halt. As yet, we have not used this enzyme which theoretically should be the enzyme of choice particularly for synthesizing long cDNAs.

Protocol 2. Synthesis of first strand cDNA

Reagents

- 5 × RT buffer: 250 mM Tris–HCl pH 8.3, 50 mM MgCl$_2$, 375 mM KCl, 50 mM DTT (store in aliquots at −20 °C, use each aliquot only once)
- 10 mM dNTP: 10 mM of each deoxynucleoside triphosphate in 10 mM Tris–HCl pH 7.5 (store at −20 °C)
- M-MLV RT (Gibco-BRL)
- [α^{32}P]dGTP or dCTP (Amersham) 3000 mCi/mmol (avoid dATP and dTTP)
- Oligo(dT)$_{12–18}$ (0.5 mg/ml)

Method

1. Heat 2–5 μg of polyadenylated RNA in 25 μl of water for 5 min at 65 °C. Chill the tube on ice.

2. Add the following in this order:
   - 5 × RT buffer   10 μl
   - 10 mM dNTPs   2.5 μl
   - oligo(dT)$_{12–18}$ 5 μl
   - RNasin   60 U
   - M-MLV RT   500 U

   Adjust to a final volume of 50 μl with ddH$_2$O.

Protocol 2. *Continued*

3. Mix and transfer 10 µl of the reaction mix to a tube containing 1 µCi of [α³²P]dGTP or dCTP. This radioactive tube is the chase reaction and will be used to determine the yield of the first strand cDNA reaction (*Protocol 4*). Incubate both reactions at 37 °C for 1 h. Place both tubes on ice and add 1 µl of 0.25 M EDTA to the labelling reaction.

3.3.2 Second strand cDNA synthesis

The preferred method for second strand synthesis is to use the mRNA in the cDNA–mRNA hybrid (the product of the first strand reaction) as the primer for *E. coli* DNA polymerase I. RNase H is added to produce gaps in the mRNA thus generating a set of primers along the length of the first strand cDNA. These RNA primers are then replaced by DNA (5). This method has almost completely superceded the use of hairpin priming followed by nuclease S1 digestion of the loop. The second strand reaction is detailed in *Protocol 3*. This step is usually very efficient, approaching 100%.

3.3.3 Blunt-ending the cDNA

The 'ragged ends' on the cDNA have to be made flush before any linkers/adaptors can be ligated on. This is carried out using T4 DNA polymerase according to *Protocol 3*, step 3. It is not straightforward to assay this step and it is easiest to assume that it has worked. Should the linkered cDNA fail to ligate into the vector, then it is quite likely that a problem has occurred at this point.

Protocol 3. Second strand synthesis and blunt-ending of cDNA

Reagents

- First strand reaction mix
- 10 × second strand buffer: 170 mM Tris–HCl pH 8.3, 50 mM MgCl₂, 900 mM KCl, 40 mM DTT
- 10 mM dNTP
- [α³²P]dGTP or dCTP (Amersham) 3000 mCi/mmol
- *E. coli* DNA polymerase (Boehringer)
- *E. coli* RNase H (Boehringer)
- T4 DNA polymerase
- Phenol equilibrated in TE buffer: 10 mM Tris–HCl pH 8.3, 1 mM EDTA
- Chloroform
- 7.5 M ammonium acetate
- 0.25 M EDTA pH 8.0

Method

1. To the first strand cDNA (40 µl) add in the following order:
   - 10 × second strand buffer 25 µl
   - 10 mM dNTP 5 µl
   - [α³²P]dGTP or dCTP 10 µCi

- *E. coli* DNA pol I   80 U
- *E. coli* RNase H   5 U

Adjust to a final volume of 250 µl with ddH$_2$O.

2. Incubate the reaction for 2 h at 16°C.

3. Heat denature the reaction mixture at 70°C for 10 min, and cool on ice. Add 5 U of T4 DNA polymerase and incubate the reaction at 37°C for 30 min (do not exceed this time).

4. Remove two 10 µl aliquots to determine the incorporated radioactivity (see *Protocol 4*), and then add 20 µl 0.25 M EDTA to the rest of the reaction.

5. Extract the reaction with an equal volume of phenol/chloroform. Flick the tube to emulsify, then separate the phases by centrifugation in a microcentrifuge for 3 min. Transfer the upper aqueous phase to a fresh tube.

6. Precipitate the cDNA by adding 0.5 vol. 7.5 M ammonium acetate and 2.5 vol. ethanol, and centrifuge at room temperature for 15 min.

7. Resuspend the cDNA in 200 µl of ddH$_2$O and re-precipitate with ammonium acetate and ethanol. Wash the pellet with 75% ethanol and air dry briefly. Resuspend the cDNA in 20 µl of ddH$_2$O. After the yield of cDNA has been determined, adjust the volume to give 1 µg of cDNA per 20 µl of TE buffer.

3.3.4 Assessing cDNA reactions

The first and second strand reactions should be assayed in two ways:

- Calculate the yield of cDNA by measuring the incorporation of radiolabel.
- Determine the size of the cDNA by alkaline agarose gel electrophoresis.

A sample calculation of the yields of first and second strand cDNA is given in *Protocol 4*. Typical conversion of mRNA to cDNA is around 30% by weight but could be in the range 10–50%. The products of the first and second strand synthesis reactions should be electrophoresed on a denaturing (alkaline) agarose gel to assess the size of the cDNA (*Protocol 5*). An example of such an analysis is shown in *Figure 2*. A number of distinct bands can be seen because the source of the RNA was lactating mouse mammary gland which synthesizes very large amounts of specific milk protein mRNAs. The sizes of the major bands are the same for both first and second strands, demonstrating that complete copying of the first strand has taken place and that no appreciable hairpinning has occurred since no 'double-sized' bands can be seen. However, for the majority of cells and tissues, there will be few bands and a smear from 400 bp to about 8 kb will be seen.

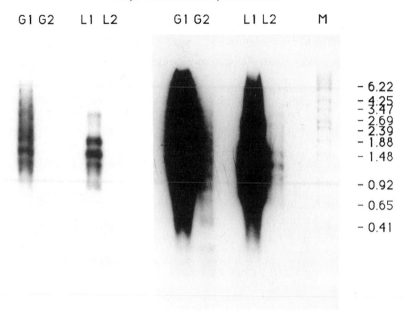

G1 G2   L1 L2   G1 G2   L1 L2   M

- 6.22
- 4.25
- 3.47
- 2.69
- 2.39
- 1.88
- 1.48

- 0.92

- 0.65

- 0.41

Figure 2. Alkaline agarose gel analysis of first and second strand cDNAs synthesized from 15 day-pregnant and 11 day-lactating mammary gland mRNA. Aliquots of first and second strand reactions were prepared as described in *Protocol 5*. In this experiment the second strand was made about ten times more radioactive than the first strand. Lanes L2 and L1 are first and second strand reactions, respectively, from lactating mammary gland RNA, while G2 and G1 are the first and second strand products from pregnant mammary gland RNA. The panel on the *right* is a longer exposure. The markers (lane M) are 20 μl of ^{35}S-labelled DNA markers (Amersham). A number of discrete bands are visible, particularly in the lactating cDNA tracks. This reflects the high abundance of milk protein mRNAs. Such a pattern would not usually be seen with cDNAs prepared from other tissues.

Protocol 4. Estimation of cDNA yield

Reagents

- First strand chase reaction
- Aliquots from the second strand reaction
- 10% (w/v) TCA solution in ddH$_2$O
- Whatman GF/C glass fibre filters

- TE buffer: 10 mM Tris–HCl pH 8.3, 1 mM EDTA
- Standard scintillant

Method

1. To the 10 μl first strand chase reaction, add 90 μl TE buffer and mix well. Spot two 5 μl aliquots from the diluted reaction on to separate GF/C filters. Allow to air dry. Then place one of the filters into a scintillation vial. The other will be treated with TCA (step **3**). Note that it is important

to mark each filter carefully before applying the radioactive aliquots so that each filter can be identified correctly should they get mixed up.

2. Spot one of the 10 µl aliquots from the second strand reaction directly on to a GF/C filter. Allow to air dry. This filter will be treated with TCA. Dilute the second 10 µl aliquot with 90 µl of TE buffer, mix well, and then spot 5 µl of this diluted mixture on to a separate GF/C filter. Allow to air dry and place in a scintillation vial.

3. Wash the GF/C filters containing the first and second strand aliquots with ice-cold TCA three times, for 5 min in about 50 ml. Wash the filters once in 95% ethanol for 5 min at room temperature, and then allow them to dry completely (use a heat lamp or hair drier). Place filters in scintillation vials.

4. Add standard scintillant (1–2 ml) to each vial and count each vial (three times), for 1 min with a scintillation counter. Alternatively, omit scintillant and count for 1 min on the tritium channel (Cerenkov counting).

5. Determining first strand yield.
   Yield = (TCA precipitated c.p.m./total c.p.m.) × 6.6.

6. Determining second strand yield.
   Yield = (TCA precipitated c.p.m./total c.p.m.) × 4.62.

Protocol 5. Alkaline agarose gel analysis of first and second strand cDNA

Reagents

- Remainder of the diluted first strand labelling reaction
- Blunt-ended cDNA from *Protocol 3*
- 10 × gel solution: 30 mM NaCl, 2 mM EDTA
- 10 × alkali gel electrophoresis solution: 0.3 M NaOH, 20 mM EDTA
- 2 × loading dye: 20 mM NaOH, 20% glycerol, 0.02% bromophenol blue
- Size markers (^{35}S-labelled DNA, Amersham)
- Agarose
- 10% acetic acid
- TE buffer: 10 mM Tris–HCl pH 8.3, 1 mM EDTA
- 7.5 M ammonium acetate

Method

1. Prepare a 1.2% alkaline agarose gel (in 1 × gel solution) and cover the gel with 1 × alkali gel electrophoresis solution. Allow the gel to soak for 1 h prior to electrophoresis.

2. Precipitate the labelled first strand cDNA from the diluted mixture (*Protocol 4*, step 1) by adding 45 µl of 7.5 M ammonium acetate and 225 µl of ethanol. Collect the DNA by microcentrifugation for 10 min at room temperature. Wash the pellet with 75% ethanol and briefly air dry. Resuspend the cDNA in 5 µl TE buffer.

Protocol 5. *Continued*

3. Dilute 2 μl of the second strand cDNA (*Protocol 3*, step 6) to 5 μl with TE buffer. Add an equal volume of loading dye to the first and second strand aliquots of cDNA and to the DNA size markers. Incubate for 5 min at room temperature to denature the DNA.

4. Electrophorese the samples at 150 mA for 3 h recirculating the electrophoresis solution or until the dye reaches the bottom of the gel. Ensure that the gel does not become too hot.

5. Fix the gel in 10% acetic acid for 30 min then dry on to 3 MM filter paper using a gel drier under vacuum with the heat off (otherwise the agarose will melt).

6. Autoradiograph the dried gel. An overnight exposure should be sufficient for a preliminary analysis.

3.4 Preparing cDNA for cloning

3.4.1 Addition of linkers or adaptors

Restriction sites have to be added to the blunt-ended cDNA so that it can be cloned into the desired vector. These restriction sites can be added by ligating either linkers or adaptors on to the cDNA. A linker contains a restriction site in the middle of a double-stranded oligomer, and therefore has to be cut before the cDNA insert can be cloned. This requires methylation of the appropriate restriction site within the cDNA prior to ligation of the linkers. During the ligation, linkers form concatemers on to the end of the cDNA inserts. Excess linkers are removed first by a restriction digest and then separated away from the cDNA by a size fractionation step. Adaptors are made up of two different oligonucleotides which are of different lengths. Of these, only one oligomer (usually the shorter) is phosphorylated, thereby preventing the adaptors from forming concatemers during the ligation and ensuring they only ligate on to the blunt-ended cDNA. The adaptors which have not ligated are removed by a size separation step and the adapted cDNA is phosphorylated by a kinase reaction prior to cloning. An advantage of adaptors is that they often contain more than one restriction site which can make subcloning of the insert at a later stage more convenient. If directional cloning is desired one site will have already been added during cDNA synthesis by the oligo (dT) primer and this will have to be digested to create a 'sticky' end. It is still necessary, however, to add an adaptor or linker to the 5' end of the cDNA insert before cloning.

As an alternative to linker–adaptor ligation, the cDNA can be synthesized on the vector, as originally described by Okayama and Berg (10). This method is difficult to carry out because of the homopolymeric tailing step. However, a vector primed kit is available from United States Biochemical (cDNA ClonStruct™ Kit). The homopolymeric tails can cause problems during DNA sequencing because DNA polymerases 'stutter' when copying large tracts of a single nucleotide rendering the sequence unreadable. Despite these drawbacks, the advantages of this strategy are threefold:

(a) The ligation is intramolecular thereby increasing the efficiency of the ligation and consequently the size of the library.

(b) The first strand cDNA is treated with terminal transferase to add a homopolymeric tail. Since this enzyme is most efficient in adding nucleotides to blunt-ends or overhanging 3' ends, partial cDNA copies (which result in a 3' recessed end) are not efficiently tailed. This results in a selection in favour of full-length clones.

(c) Cloning is directional.

The use of adaptors is probably the method of choice at present as it is quicker and avoids the methylation steps. One other way to avoid methylation is to incorporate a methylated derivative of one of the deoxynucleotides into the first strand cDNA synthesis reaction. The ds DNA will then be resistant to cleavage with most restriction enzymes. This step is used in the Stratagene cDNA cloning kit and is well-worth considering as an alternative to methylation of specific restriction endonulcease sites. In *Protocols 6* and *7* we describe how to prepare cDNA for cloning by using either linkers or adaptors, respectively. The protocols describe the use of *EcoRI* linkers/ adaptors but are generally applicable. A good range of adaptors and linkers are available from New England Biolabs.

Protocol 6. Preparing cDNA for cloning using linkers

Reagents

- S-adenosyl-L-methionine (SAM (New England Biolabs) supplied with *Eco*RI methylase): dilute 1 μl of 30 mM SAM with 37.5 μl of ddH$_2$O
- 5 × methylase buffer: 500 mM Tris–HCl pH 8.0, 500 mM NaCl, 5 mM EDTA pH 8.0

- Bovine serum albumin (BSA), 2 mg/ml, nuclease free (Boehringer)
- *Eco*RI methylase (New England Biolabs)

A. *Methylation of cDNA*

1. Add to 20 μl of cDNA (1 μgl):
   - SAM   3 μl
   - 5 × methylase buffer   3 μl
   - BSA   1.5 μl
   - *Eco*RI methylase   20 U

   Adjust to a final volume of 30 μl with ddH$_2$O.

2. Incubate at 37 °C for 60 min. Heat denature the methylase at 70 °C for 10 min, cool on ice, and proceed to part B.

Protocol 6. *Continued*

Reagents

- 10 × kinase buffer: 100 mM Tris–HCl pH 7.5, 100 mM MgCl$_2$, 50 mM DTT, 10 mM ATP (store in aliquots at −20°C, and use once only)
- *Eco*RI linkers (NEB) d(CGAATTCCG) (if phosphorylated linkers are purchased, step **1** may be omitted)
- [γ^{32}P]ATP (Amersham PB10205) specific activity 3000 mCi/mmol)
- T4 polynucleotide kinase (Boehringer)
- T4 DNA ligase (Boehringer)

B. *Addition of linkers to cDNA*

1. The linkers should first be phosphorylated. Set up a 50 μl reaction containing 5 μg of linkers and add the following:
   - 10 × kinase buffer   5 μl
   - [γ^{32}P]ATP   0.5 μl (5 μCi)
   - T4 polynucleotide kinase   30 U

   Adjust to a final volume of 50 μl with ddH$_2$O.
   Incubate for 1 h at 37°C, then heat denature the kinase at 70°C for 10 min, and freeze 5 μl aliquots at −20°C.

2. To the 30 μl of cDNA from the methylation reaction (part A), add:
   - 10 × kinase buffer   2 μl
   - kinased linkers   5 μl (500 ng)
   - T4 DNA ligase   10 Weiss units

   Adjust to a final volume of 50 μl with ddH$_2$O and incubate at 14°C overnight.

3. Heat inactivate the ligase by incubating at 70°C for 10 min. Centrifuge briefly to collect droplets of condensation. Proceed to part C.

Reagents

- *Eco*RI restriction buffer (high salt; Boehringer)
- *Eco*RI (high concentration; Boehringer)

C. *Removal of excess linkers*

1. To the 50 μl of linker ligation add:
   - *Eco*RI buffer   10 μl
   - *Eco*RI   50 U

   Adjust to a final volume of 100 μl with ddH$_2$O.

2. Incubate at 37°C for 1 h, then add another 50 U of *Eco*RI to the reaction, and incubate for another hour at 37°C. Proceed to size fractionation, *Protocol 8*.

Protocol 7. Preparing cDNA for cloning using adaptors

Reagents

- 10 × ligase buffer: 100 mM Tris–HCl pH 7.5, 100 mM MgCl$_2$, 50 mM DTT, 10 mM ATP (store in aliquots at −20°C, and use once only)
- *Eco*RI adaptors from New England Biolabs, e.g. unphosphorylated EcoRI–*Xma*l d(AATTCGAACCCCTTCG) oligonucleotide A, and its complement phosphorylated

- *Xma*l d(pCGAAGGGGTTCG) oligonucleotide B
- 3 M NaCl
- 1 M Tris–HCl pH 8.0
- 0.1 M EDTA pH 8.0
- T4 DNA ligase (Boehringer)
- T4 polynucleotide kinase (Boehringer)

Method

1. To anneal adaptors set-up the following reaction:
   - oligonucleotide A   10 μg (unphosphorylated)
   - oligonucleotide B   10 μg (phosphorylated)
   - 3 M NaCl   6.7 μl
   - 1 M Tris–HCl pH 8.0   2μl
   - 0.1 M EDTA pH 8.0   2 μl

   Adjust to a final volume of 200 μl with ddH$_2$O.
   Place in a 100 ml beaker of boiling water and allow the beaker to cool to room temperature. Aliquot and store at −20°C.

2. To 20 μl of cDNA (1 μg) add the following:
   - 10 × ligase buffer   3 μl
   - *Eco*RI adaptors   1 μg
   - T4 DNA ligase   5 Weiss U

   Adjust to a final volume of 30 μl with ddH$_2$O.
   Incubate at 14°C overnight. Carry out size fractionation of the cDNA (*Protocol 8*) and then continue on to step **3**.

3. To phosphorylate the size fractionated, adaptor ligated cDNA add the following to 20 μl of cDNA:
   - 10 × ligase buffer   3 μl
   - T4 polynucleotide kinase   30 U

   Adjust to a final volume of 30 μl with ddH$_2$O.
   Incubate at 37°C for 30 min. Then extract the reaction with phenol/chloroform, and then chloroform (see *Protocol 3*, step 5).

3.4.2 Size fractionation of cDNA

After ligating the linkers or adaptors to the cDNA (*Protocols 6* or *7*), excess linkers or adaptors are removed by chromatography on Ultrogel AcA34, or Biogel A50m, or Sephacryl spin columns. This step can also be used as a size fractionation of the cDNA itself. In *Figure 3*, aliquots of

103

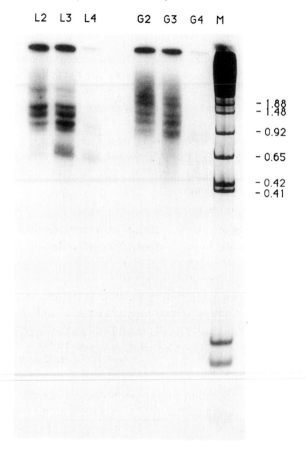

Figure 3. Analysis of fractions eluted from the size fractionation column. Aliquots (5 μl) of each fraction were electophoresed on a native 6% acrylamide gel which was then fixed in 10% acetic acid for 10 minutes and dried under vacuum before autoradiography. Lanes L2–L4 are lactating mammary gland cDNA samples, and G2–G4 are pregnant mammary gland cDNAs. Fractions containing the longest cDNAs were pooled. Markers (lane M) are ^{35}S-labelled DNA markers (Amersham). Note that the relative mobility of the marker fragments is different on the native gel compared to the alkaline agarose gel.

fractions obtained from a Sephacryl column were separated on a 6% poly-acrylamide gel, the gel fixed, dried, and then autoradiographed. Short cDNAs can be excluded by pooling only the first few fractions.

Protocol 8. Size fractionation of cDNA

Reagents

• Sephacryl spin column (Stratagene)
• 10 × STE: 100 mM NaCl, 10 mM Tris–HCl pH 7.5, 1 mM EDTA

Method

1. Add 30 μl ddH$_2$O and 5 μl 10 × STE to the cDNA from *Protocol 6C*, step 1 or *Protocol 7*, step 3. Prepare the spin column according to the manufacturer's recommendations and place the column tip in a microcentrifuge tube which has had the cap cut off. Load the cDNA on to the column using a Pipetman, then place the column assembly into a plastic tube.

2. Centrifuge at 2000 r.p.m. in a Sorvall HB4 swing-out rotor (or its equivalent) for 2 min. Remove the microcentrifuge tube with the first eluted fraction. Replace with a fresh tube, load 60 μl of 1 × STE on to the column, and spin as before.

3. Repeat this step three more times.

4. Run 5 μl aliquots of each fraction on a 6% non-denaturing acrylamide gel along with suitable size markers. When the electrophoresis is complete, fix the gel in 10% acetic acid for 10 min then dry on a gel drier, and autoradiograph overnight at −70 °C.

5. Pool the fractions containing the cDNA of the desired size. Add an equal volume of phenol/chloroform (1:1), mix, and spin in a microcentrifuge for 2 min. Remove the aqueous phase and extract with chloroform. Ethanol precipitate the cDNA by adding 2.5 vol. ethanol (there is no need to add salt in addition to the NaCl in the STE). Put at −20 °C overnight.

6. Pellet the cDNA by microcentrifugation at 4 °C for 60 min. Carefully remove the supernatant. Check that all of the radioactivity is contained in the pellet. If any counts can be detected in the ethanol, re-centrifuge as above.

7. Spin briefly to collect droplets of ethanol, carefully remove by pipetting, and air dry the pellet. Resuspend in 10 μl ddH$_2$O and freeze at −20 °C until required. Note that cDNA ligated to adaptors has to be phosphorylated (*Protocol 7*, step 3) before proceeding to the cloning stage.

3.5 Cloning of cDNA

3.5.1 Preparation of vector DNA and cDNA ligation

The vector DNA should be digested with the appropriate restriction enzyme(s) and dephosphorylated with alkaline phosphatase. The dephosphorylation step greatly reduces the number of self-ligated vector clones in the cDNA library. Most commercial vectors mentioned in *Table 1* can be bought as pre-cut and dephosphorylated DNA thereby avoiding the vector preparation step. Procedures for vector preparation and ligation of cDNA are given in *Protocol 9*. This protocol assumes ligation to *Eco*RI sites in the vector. The appropriate enzyme(s) for the cloning site(s) should of course be used. The following points should be noted:

(a) The molar ratio of vector to cDNA should be at least 2:1 to reduce the chance of double cDNA inserts. This ratio may have to be determined empirically as not all cDNA will have a compatible restriction site and the average size of the insert has to be estimated. It is therefore wise to set-up several ligations with different amounts of cDNA.

(b) When using lambda vectors the samples should be quite viscous if the ligation has been successful.

(c) A control ligation of vector arms or plasmid only should be carried out. Some commercial kits provide a positive control for insert ligation (e.g. Amersham). Alternatively pUC19 DNA can be cut with, e.g. *Eco*RI, diluted to 100 ng, and used as a test insert.

Protocol 9. Digestion of vector DNA and ligation to cDNA

Reagents

- Vector DNA of choice
- 10 × *Eco*RI buffer
- *Eco*RI (high concentration)
- Phenol/chloroform (1 : 1) (equilibrated)
- Ethanol
- TE: 10 mM Tris–HCl pH 8.0, 1 mM EDTA pH 8.0

- 10 × ligase buffer: 100 mM Tris–HCl pH 7.5, 100 mM MgCl$_2$, 50 mM DTT, 10 mM ATP (store in aliquots at −20 °C and use once only)
- Calf intestinal alkaline phosphatase (1 U/μl, DNase-free; Boehringer)
- T4 DNA ligase (Boehringer)

A. *Digestion of vector DNA*

1. Digest 20 μg of vector DNA in a 100 μl reaction by adding the following:
   - 10 × *Eco*RI buffer   10 μl
   - *Eco*RI   200 U

   Adjust to a final volume of 100 μl with ddH$_2$O.
   Incubate at 37 °C for 2–3 h.

2. Add 1 U of alkaline phosphatase and incubate at 37 °C for 30 min.

3. Extract twice with 100 μl of phenol/chloroform (1:1), and then once with chloroform (see *Protocol 8*, step 6). Add 2 vol. of ethanol (at room temperature) to the aqueous phase to precipitate the vector DNA. Pipette solutions of lambda vector DNA with care as it is prone to shearing.

4. Pellet the DNA by centrifugation, discard the supernatant, re-centrifuge briefly, and pipette off any remaining ethanol. Air dry briefly, ensuring that the DNA does not completely dry out or it may be difficult to resuspend.

5. Add 10 μl of TE and gently resuspend the DNA.

B. *Ligation of cDNA to vector*

1. The 10 μl ligation reaction for 100 ng of cDNA should contain:

- resuspended cDNA   100 ng[a]
- 10 × ligase buffer   1 μl
- vector arms   5 μg[a]
- T4 DNA ligase   2 Weiss units

2. Incubate overnight at 14°C. The ligated cDNA/vector will be viscous due to the formation of concatemers. It is important that shear forces are kept to a minimum. Consequently the ligation reaction should be stored at 4°C (not frozen) until packaged. Mechanical shearing can be reduced by cutting the point off a yellow tip before pipetting and mixing the solution gently.

---

[a] These amounts aim to give a 2 : 1 molar ratio of cDNA to lambda vector, assuming a mean size of 1 kb for the cDNA. Ideally, the DNA concentration in the ligation reaction will be 1 mg/ml. If the volume is too high, the cDNA and vector arms can be co-precipitated. Aim for a 1 : 1 molar ratio of cDNA to plasmid vectors, i.e. approximately 200 ng of a 2.5 kb plasmid to 100–150 ng of cDNA with an average size of 1.5–2 kb. Ligate as described in the protocol for lambda vectors. Mechanical shearing is less of a problem with plasmid vectors.

3.5.2 Packaging and plating of recombinant phage

The preparation of packaging extracts is time-consuming and technically demanding. If highly efficient 'home-made' extracts are not already available, then purchase of commercially-made extracts is recommended. These are available from a number of companies including Stratagene (Gigapack II Gold), and packaging efficiencies of up to 2×10^9 p.f.u./μg cDNA are claimed. The manufacturer's instructions should be followed closely as the preparation and type of extracts can vary widely. The use of packaging extracts which lack all known restriction activities (mcrA$^-$, mcrBC$^-$, mrr$^-$, mcrF$^-$, msdR$^-$) will optimize efficiency and the representative nature of the library, particularly if the cDNA is hemimethylated. The following points are important:

- DNA should be added immediately the freeze/thaw extract starts to thaw
- do not introduce air bubbles while mixing
- *in vitro* packaged phage are not stable and should be plated out within a day.

Once packaged, phage should be plated out according to *Protocol 10A*. It is sensible to package an aliquot of the ligation only. The following day, the remainder of the ligation can be packaged if the efficiency of packaging is of the order of 1×10^6 to 1×10^7 p.f.u./μg vector arms.

Protocol 10. Plating of recombinant phage and amplification of library

Reagents

- Suitable plating cells (at $OD_{600} = 0.5$)
- NZY agar plates: 10 g NZ amine, 5 g NaCl, 5 g Bacto yeast extract (Difco), 1 g casamino acids, 2 g $MgSO_4.7H_2O$, 15 g Bacto agar (Difco) in 1 litre, adjust pH to 7.0 with NaOH, and autoclave
- NZY top agarose: as NZY agar plates except replace Bacto agar with 7 g of agarose
- SM phage buffer: 10 mM Tris HCl pH 7.5, 10 mM $MgSO_4$, 0.01% gelatin

A. *Plating*

1. The packaged DNA should be in a final volume of 500 μl of phage dilution buffer per packaging reaction. To determine the packaging efficiency, mix both 1 μl of stock and 1 μl of a 10 × dilution of the stock with 200 μl of plating cells.

2. Pre-incubate the plating bacteria and phage at 37 °C for 15 min, then add 2.5 ml of top agarose (at 48 °C) and pour immediately on to pre-warmed NZY agar plates.

3. Incubate overnight at 37 °C with the plates inverted to prevent condensation dripping on to the surface.

4. Plate the remainder of the packaging reaction on 140 mm dishes. Use 600 μl of plating cells and approximately 5×10^4 phage per plate. Incubate at 37 °C for 15 min to allow the phage to adsorb to the bacteria, then add 7ml top agarose (at 48 °C) and pour quickly and evenly (try to prevent bubbles forming) on to pre-warmed NZY agar plates.

5. Incubate at 37 °C for at least 5 h but no more than 8 h or the plaques will become too large.

B. *Pooled plate lysate*

1. Overlay each of the plates with 10 ml of SM phage buffer and rock or shake gently at 4 °C overnight.

2. The next day, pipette off the bacteriophage suspension and collect in sterile 50 ml polypropylene tubes. Add chloroform to 5% and mix. Incubate at room temperature for 15 min.

3. Pellet the cell debris by centrifugation for 10 min at 2000 r.p.m. in a bench-top centrifuge.

4. Carefully recover the supernatants, pool in a sterile glass bottle, add chloroform to 0.3% and mix.

5. Aliquot and store at 4 °C in chloroform resistant tubes or bottles.

6. Small aliquots can also be stored at −70 °C in 7% DMSO. (Caution: DMSO is harmful.)

The following points should be noted:

(a) Always use freshly prepared cells when plating a library.

(b) It is imperative that the correct cells are used for the particular cloning vector being used.

(c) It is particularly important that large plates are dry and pre-warmed.

(d) As an alternative to NZY media LB media can be used. LB media is 10 g Bacto tryptone (Difco), 5 g Bacto yeast extract (Difco), 5 g NaCl, dissolved in 1 litre and adjusted pH to 7.2 with NaOH, and then autoclaved. LB agar is LB media with 10 g of Bacto agar (Difco) added.

3.5.3 Amplification and storage

For most purposes, it is best to amplify the library for storage, however, for some applications it is preferable to screen unamplified libraries. This situation is encountered when screening expression libraries where the clone of interest may express a protein which is toxic to the host cell or when the cDNA from the RNA of interest is unstable within the vector. Refer to *Protocol 10* for the amplification and storage of the library. The amplified library will be in a large volume of storage buffer and it is probably not necessary to keep it all. Library phage can be concentrated by CsCl density gradient banding (6).

3.5.4 Plating of recombinant plasmid

Recombinant plasmids should be transfected into suitable host cells by electroporation. Transfection of a plasmid-based cDNA library by electroporation is described in ref. 11. Electrocompetent cells are difficult and time-consuming to make, but are available from a number of commercial companies including Bio-Rad. These cells have minimum transfection efficiencies of at least 10^9 c.f.u./μg. Electroporation of an actual cDNA/plasmid ligation reaction may give lower transformation efficiencies by two to three orders of magnitude, i.e. 10^6–10^7 c.f.u./μg. To obtain efficient transformation may require the optimal pulse time to be determined empirically. Bacteria should be plated out according to the supplier's instructions. Do not allow the colonies to become too large as this will hinder identification of positive colonies in the screening procedure.

3.5.5 Assessing the quality of the library

A number of preliminary checks should be carried out to assess the quality of the library. These include:

● Estimation of the number of recombinants.

● Analysis of the insert in randomly picked plaques or colonies.

● Screening with an abundant sequence probe (e.g. β-actin).

109

The ratio of recombinants to non-recombinants can be determined in a number of ways depending on the type of cloning vector used. For λgt10, plaque morphology (clear versus turbid) is the indicator. For vectors such as λgt11 and λZAP II, a blue/white phenotype will be observed when plated with the appropriate cells on the chromogenic indicator Xgal. The ratio of blue/white (non-recombinant/recombinant) should be taken into account when calculating the complexity of the library. Recombination frequencies should approach 99% to avoid screening an excessive number of plaques. The average size of the inserts can be seen by electrophoresis of digested DNA purified from randomly picked plaques or colonies with the enzyme(s) used to cut the vector for ligation. This will also confirm that the clones are recombinant. Ten clones will be sufficient, and if all the inserts from the randomly picked clones are short, the chances of obtaining a full-length cDNA clone are low. Hybridizing a sample of the library with a probe for a highly expressed RNA gives useful information on the representative nature of the library and confirms the frequency of recombination. *Figure 4* shows the result of hybridizing 10 000 plaques from the mammary cDNA library with a β-casein probe. Approximately 14% of the plaques are positive, which agrees well with the estimated level of β-casein transcripts in mammary gland mRNA.

4. Library screening—standard procedures

The following sections deal with methods for isolating cDNA clones by hybridization with nucleic acid probes. Libraries constructed in expression vectors such as λgt11 and λZAP II can be induced to express fusion proteins which can then be detected by specific antibody binding. Methods for screening expression libraries with antibodies are discussed in *DNA cloning: a practical approach* Vol. 2.

4.1 Screening a library with DNA probes

Procedures for preparing plating cells, plating out a library, and taking replica filters are given elsewhere (6). A variety of probes can be used including large DNA insert fragments from plasmids (> 200 bp), oligonucleotides from known DNA sequences, or degenerate oligonucleotides corresponding to segments of known protein sequences, and PCR amplified fragments. Preparation and hybridization of probes is described in Chapter 4. Once positive plaques have been identified, they should be subjected to two further rounds of hybridization until a plaque purified stock has been obtained. Insert cDNA should then be subcloned into a suitable plasmid vector or *in vivo* excised, and the cDNA sequenced as a first step in validating the clone.

Apart from this relatively simple procedure, there are a number of other

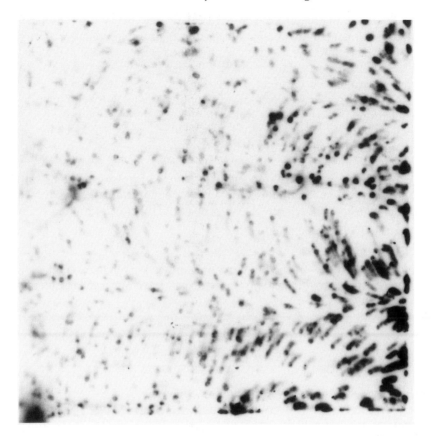

Figure 4. Test hybridization of an aliquot of the library with an abundant sequence probe. An aliquot of the lactating mammary gland cDNA library containing approximately 10^4 p.f.u. was plated out on 20 cm × 20 cm square plates, and filters prepared for hybridization to a β-casein probe, radiolabelled by random priming. Around 14% of the plaques gave a positive signal which correlates well with the estimated level of β-casein mRNA (15–20%). Note that it is difficult to lift such large filters without causing streaking of the DNA.

approaches which can be used to screen cDNA libraries. It is beyond the scope of this chapter to give detailed protocols for these. The following sections discuss some of the options and provide references for further information.

4.2 Differential screening

In some situations, the aim is to isolate a particular cDNA clone or to clone a class of factors for which no sequence information is available and no antibodies have been generated. If the gene of interest is expressed at higher levels in one cell than in a closely related cell (e.g. is induced or repressed by hormone treatment, growth conditions, etc.) then differential screening can be applied. This procedure is also used to clone transcripts which are induced in response to a particular treatment. Probes are prepared from mRNA isolated from the two different populations of cells. Radiolabelled, single-stranded cDNAs are synthesized from each of these mRNAs and hybridized to replica filters. Most plaques will hybridize equally with both probes, but some will show a clear difference in signal strength. These clones may represent differentially expressed mRNAs. Often three successive rounds of hybridization are necessary to isolate true differentially expressed clones. Genes which are involved in a range of cellular processes including differentiation have been cloned this way (12). A commercial labelling kit which produces high specific activity cDNA probes suitable for differential screening is available from Promega.

There are a number of potential problems with this technique:

(a) Plaques must be screened at very low density to facilitate the identification of differential signals. It is therefore difficult to screen large numbers of clones.

(b) Signal strength depends on the relative abundance of the mRNA. Rare RNAs will be poorly represented in the cDNA probe, resulting in a weaker hybridization signal. This makes it difficult to spot any differences between the replicas.

(c) Very low abundance mRNAs will be difficult to clone this way since they will be a very minor constituent of both the library and the probe. For any mRNA species with an abundance of less than about 0.1%, subtraction cloning is more likely to be successful.

4.3 Subtracted probes and subtraction libraries

Subtraction cloning is a much more complex procedure than differential screening. Both subtracted probes and subtracted libraries may be constructed. A number of variations on the general theme have been tried and the papers referred to below give detailed protocols. One example is to biotinylate common sequences and remove these using streptavidin (13). Another approach is the use of thiol labelling of newly transcribed RNA which can then be selected on a phenylmercury agarose (Affi-Gel 501) column (14). A magnet assisted subtraction technique has also been described (15). It should be mentioned that subtraction cloning is a difficult procedure which will not be successful in every case. A few companies produce kits for subtraction cloning (e.g. Invitrogen). The recent development of the differential display (RNA fingerprinting) procedure (discussed in Section 5.4) will possibly render differential screening and subtraction cloning redundant in most situations.

5. Specialized cDNA cloning techniques using the polymerase chain reaction (PCR)

In this section, we will discuss the various options that are available for cloning cDNAs without constructing and screening a conventional cDNA library. This includes a number of different procedures, based on PCR, whose use will be dictated by the particular circumstances. For example, it may be necessary to construct a cDNA library from very limited amounts of material, the required cDNAs may be differentially expressed in different cell types or be present at very low levels, or the cDNA of interest may be part of a family of genes. These types of cDNAs are usually more difficult to isolate and require specialized cloning techniques.

5.1 Polymerase chain reaction coupled reverse transcription (RT-PCR)

In principle, any cDNA for which limited sequence information is available, can be cloned using reverse transcription coupled with amplification by PCR. A generalized strategy for RT-PCR cloning is schematized in *Figure 5*. First strand cDNA is synthesized using oligo(dT) primers or, for ease of cloning, a primer—adaptor composed of oligo(dT) linked to a restriction enzyme recognition site. Single-stranded cDNA is then amplified by adding a second primer derived from the known sequence. Usually about 30 amplification cycles are required. Alternatively, two sequence-specific primers may be used.

The known sequence may be derived from short peptide fragment sequences from purified proteins. The amino acid sequence will generate a mixture of redundant oligonucleotides which can be used as PCR primers clone short segments of cDNA between the known amino acid sequences (16). Novel members of multigene families can also be cloned using RT-PCR. Short oligonucleotide primers can be generated from regions of high sequence conservation between known family members. For example, the sequence of functional domains in transcription factor families (DNA binding, ligand binding, dimerization) are often conserved. We, and others, have cloned novel members of the *Wnt* family of signalling molecules using RT-PCR (17). *Protocol 11* provides details of the reaction conditions for cloning *Wnt* genes by PCR. Other applications of RT-PCR include exon trapping (18) and quantitation of RNA transcripts (commercial kits are now available).

RT-PCR products should be cloned in plasmid vectors using high efficiency competent cells—electroporation is not essential. A general purpose cloning vector such as pBluescript is adequate. However there are a number of vectors which are specifically designed for cloning PCR products and these are summarized in *Table 2*.

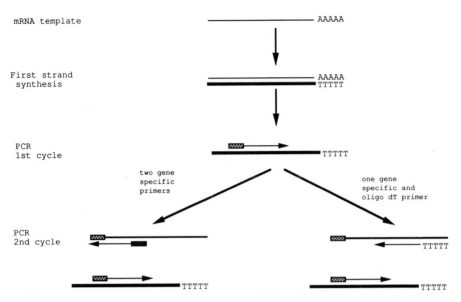

Figure 5. RT-PCR cloning of cDNA. The first cDNA strand is synthesized by reverse transcriptase in the standard way to generate a template for the first PCR cycle. A gene-specific primer (which may consist of a pool of oligonucleotides containing all the possible coding sequences for a particular tract of amino acids) is used for the specific amplification of the cDNA of interest. Note that inosine can be incorporated into the oligo-nucleotides at positions of high degeneracy. A second PCR amplification cycle, using this first strand cDNA as a template, is then carried out using either a second gene-specific primer or oligo(dT). The PCR products of this reaction can then be cloned into a suitable vector.

Protocol 11. RT-PCR cloning

Reagents

- Gene-specific primers at a concentration of 10μM, one 'sense' and the other 'anti-sense'[a]
- 2 mM dNTP (dATP, dCTP, dGTP, dTTP), each at 2 mM
- Taq DNA polymerase (Boehringer)
- 10 × reaction buffer supplied with Taq polymerase

- Poly (A[+]) RNA or total RNA
- Components for first strand cDNA synthesis (see Protocol 2)
- 10 × ligase buffer
- T4 DNA ligase

Method

1. Prepare template DNA. Using the standard reaction conditions in *Protocol 2*, synthesize first strand cDNA from either an mRNA template (1–5 μg) or 20 μg of total RNA using oligo(dT) as primer.

2. For the PCR reaction, use 2μl of the reaction mix or 10–50 ng of cDNA

114

(depending on the level at which the target mRNA is expressed, i.e. use more for rare transcripts), and add the following:

- 10 × reaction buffer   5 μl
- 5 × dNTP mix   10 μl
- primer 1   5 μl
- primer 2   5 μl
- Taq DNA polymerase   2.5 U

Adjust the final volume to 50 μl with ddH$_2$O.
Carry out the amplification in a PCR machine using the following typical conditions:[b]

- 94°C 1 min
- 55°C 1 min (vary this to suit the melting temperature of the primer)
- 72°C 1 min (use 1 min/200 bp)

25–30 amplification cycles should be carried out.

3. Carry out a final extension at 72°C for 8 min.

4. Electrophorese the reaction products in a 1–2% agarose gel (depending on expected PCR product size) with suitable size markers. Excise the specific product from the gel and use a geneclean kit (Stratagene) to remove PCR reaction components.

5. Digest the DNA with the appropriate restriction enzymes (XhoI and XbaI) and perform a second geneclean step.

6. Ligate 100 ng of the PCR product to the chosen plasmid vector as follows:

- PCR fragment   100 ng
- XhoI/XbaI cut vector   100 ng
- 10 × ligase buffer   2 μl
- T4 DNA ligase   1 Weiss unit

Adjust the final volume to 20 μl with ddH$_2$O.
Incubate overnight at 16°C.

7. Use 5 μl of the ligation reaction to transform 200 μl of the appropriate competent cells using protocols described in Chapter 1, and plate out with appropriate indicators/inducers/selective antibiotics as dictated by the choice of vector.

---

[a] Primers should be about 15–24 bases in length and have a melting temperature of around 60–80°C. It is useful to add a restriction site sequence at the 5′ end of each primer. The use of a different site on each primer allows directional cloning of the PCR product. We have used XhoI, EcoRI, and XbaI successfully.

[b] Annealing temperatures and extension times will depend on the sequence of the primers being used and the length of the fragment to be amplified. Computer programs are available to assist in determining optimal primers and conditions.

Table 2. Plasmid vectors for cloning PCR products

Name	Source	Features
p-Direct	Clontech	Directional, ligation not needed
pCR-Script™ SK(+)	Stratagene	Fast, low false-positive rate
pGEM-T	Promega	T-overhangs
pAmp™	Gibco-BRL	Directional, ligation not needed

5.2 Rapid amplification of cDNA ends (RACE)

Often cDNA clones are not full-length and it is usually necessary to isolate overlapping cDNA clones in order to derive the complete sequence. Even then, it is often not possible to determine the sequence at the extreme 5' end of a clone. This is a particular problem for large or low abundance transcripts. The RACE technique was originally described by Frohman (19) and was devised to overcome the problem of cloning the 5' ends but can also be applied to cloning of 3' ends of cDNAs. RACE uses PCR to amplify a region of a particular cDNA from a single point within the transcript (therefore a short stretch of sequence must be known) and either the 3' or 5' end of this clone. The 3' and 5' end primer sequences are derived either from the vector in which the cDNA template is cloned or from adaptors ligated on to the cDNA. A detailed RACE protocol is provided in ref. 20, and although this procedure does work it can be technically demanding with many clones potentially having to be characterized. A modification of RACE which eliminates the need for homopolymeric tailing is provided by the SLIC method (21). A kit based on this approach is produced by Clontech. Both 5' and 3' RACE systems are also produced by Gibco-BRL.

5.3 Small amounts of mRNA

It is sometimes difficult to obtain sufficient quantities of RNA to construct a conventional cDNA library. Over 5000 mouse eggs were required to construct a cDNA library, for example (11). New techniques are available now for purifying RNA from small amounts of cells or tissues (REX kit from United States Biochemical, GlassMAX™ RNA spin cartridge system from Gibco-BRL). In this situation, amplification of the first cDNA strand by PCR can be carried out by adding a homopolymeric tail to the 3' end of the cDNA (e.g. G residues) or by adding adaptors to the cDNA. Oligo(dT) and oligo(dC) or adaptor primers can then be used for PCR amplification of the cDNA (22). One problem with this technique is that a proportion of the cDNAs will contain errors, thus several independent clones will need to be isolated and sequenced. The SuperScript™ pre-amplification system (Gibco-BRL) is designed for the synthesis of first strand cDNA from limited amounts of RNA using PCR as is the cDNA Cycle™ Kit from Invitrogen.

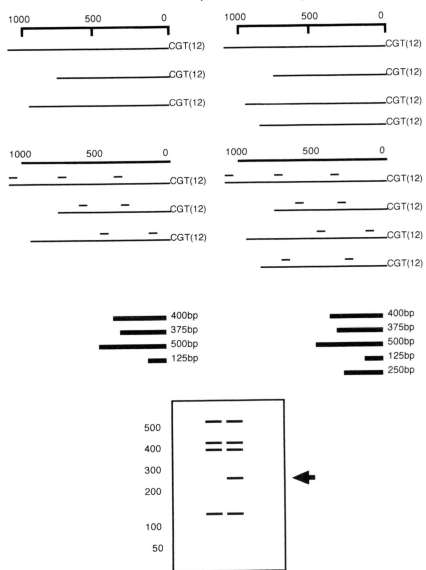

Figure 6. cDNA cloning by differential display (RNA fingerprinting). The panel on the *left* represents mRNAs from one cell type or tissue while that on the *right* represents transcripts from a closely related cell or tissue. Only one primer, CG(T)$_{12}$ is illustrated. This will anneal to a number of mRNAs and prime the synthesis of single-stranded cDNAs which will vary in length. A fourth transcript is shown in the *right* panel to represent a 'difference' between the two sources of mRNA. Random priming and synthesis of the complementary cDNA strand will produce a large number of short fragments. When these cDNAs are resolved by electrophoresis on an acrylamide gel, the fragments which are specific for the differentially expressed mRNA will be displayed as additional bands. These can be excised from the gel and cloned.

5.4 Differential display (RNA fingerprinting)

This procedure was devised by Liang and Pardee (23) and is a simpler yet more powerful way to clone differentially expressed genes. The method utilizes PCR and two sets of oligonucleotide primers and is outlined in *Figure 6*. One set consists of short primers of arbitrary sequence and the other of oligo(dT) primers with two additional 3′ bases which become anchored at the poly(A) tail of the mRNA. Since the primers are entirely random once they have been synthesized they can be used for other experiments. Improvements to the original protocol have recently been published (24, 25). This procedure is relatively untested but may prove to be the method of choice for cloning differentially expressed genes.

There are a number of other ways in which cDNAs can be cloned. An olfactory neuronal transcription factor cDNA was isolated by genetic selection in yeast by cloning cDNA into a yeast expression vector (26). A number of other ingenious procedures have been devised. The Matchmaker™ system from Clontech uses a GAL4 activation domain hybrid cloning vector to clone interacting proteins. The sophistication of cDNA cloning vectors is rapidly increasing!

Acknowledgements

We would like to thank Diane Redmond for the RT-PCR protocol, Katrina Gordon for *Figure 4*, and Mike Clinton for *Figure 6*. C.J.W. is supported by an AFRC post-doctoral fellowship and J.D. by a Wellcome Trust/Health Research Council of New Zealand Overseas Fellowship.

References

1. Rabbitts, T. H. (1976). *Nature*, **260**, 221.
2. Legon, S., Glover, D. M., Hughes, J., Lowry, P. J., Rigby, P. W. J., and Watson, C. J. (1982). *Nucleic Acids Res.*, **10**, 7905.
3. Clarke, L. and Carbon, J. (1976). *Cell*, **9**, 91.
4. Williams, J. G. (1977). *Cell*, **17**, 903.
5. Watson, C. J. and Jackson, J. F. (1985). In *DNA cloning: a practical approach* (ed. D. M. Glover), Vol. 1, pp. 79–88. IRL Press, Oxford.
6. Sambrook, J., Fritsh, E. F., and Maniatis, T. (ed.) (1989). *Molecular cloning: a laboratory manual* (2nd edn). Cold Spring Harbor Laboratory Press, New York.
7. Davis, R. W., Botstein, D., and Roth, J. R. (1980). In *Advanced bacterial genetics*. Cold Spring Harbor Laboratory, Cold Spring Harbor, New York.
8. Murphy, A. and Schimke, R. (1991). *Nucleic Acids Res.*, **19**, 3403.
9. Aviv, H. and Leder, P. (1972). *Proc. Natl. Acad. Sci. USA*, **69**, 1408.
10. Okayama, H. and Berg, P. (1982). *Mol. Cell. Biol.*, **2**, 161.
11. Rothstein, J. L., Johnson, D., DeLoia, J. A., Skowronski, J., Solter, D., and Knowles, B. (1992). *Genes Dev.*, **6**, 1190.

12. Gorman, C. M., Lane, D. P. L., Watson, C. J., and Rigby, P. W. J. (1985). *Cold Spring Harbor Symp. Quant. Biol.*, **50**, 701.
13. Owens, G. R., Hahn, W. E., and Cohen, J. J. (1991). *Mol. Cell. Biol.*, **11**, 4177.
14. Beadling, C., Johnson, K. W., and Smith, K. A. (1993). *Proc. Natl. Acad. Sci., USA*, **90**, 2719.
15. Schraml, P., Shipman, R., Stulz, P., and Ludwig, C. U. (1993). *Trends Genet.*, **9**, 71.
16. Ghosh, S., Gifford, A. M., Riviere, L. R., Tempst, P., Nolan, G. P., and Baltimore, D. (1990). *Cell*, **62**, 1019.
17. Gavin, B. J., McMahon, J. A., and McMahon, A. P. (1990). *Genes Dev.*, **4**, 2319.
18. Buckler, A. J., Chang, D. D., Graw, S. L., Brook, D., Haber, D. A., Sharp, P. A., and Housman, D. E. (1991). *Proc. Natl. Acad. Sci. USA*, **88**, 4005.
19. Frohman, M. A., Dush, M. K., and Martin, G. R. (1988). *Proc. Natl. Acad. Sci. USA*, **85**, 8998.
20. Frohman, M. A. (1990). In *PCR protocols: a guide to methods and applications.* (ed. M. A. Innis, D. M. Gelfaud, J. J. Sninsky, and T. J. White), pp. 28–38. Academic Press, Inc., New York.
21. Dumas, J. B., Edwards, M., Delort, J., and Mallet, J. (1991). *Nucleic Acids Res.*, **19**, 5227.
22. Belyavsky, A., Vinogradava, T., and Rajewsky, K. (1989). *Nucleic Acids Res.*, **17**, 2919.
23. Liang, P. and Pardee, A. B. (1992). *Science*, **257**, 967.
24. Bauer, D., Muller, H., Reich, J., Riedel, H., Ahrenkiel, V., Warthoe, P., and Strauss, M. (1993). *Nucleic Acids Res.*, **21**, 4272.
25. Liang, P., Averbaukh, L., and Pardee, A. B. (1993). *Nucleic Acids Res.*, **21**, 3269.
26. Wang, M. M. and Reed, R. R. (1993). *Nature*, **364**, 121.

4

Making nucleic acid probes

LUKE ALPHEY and HUW D. PARRY

1. Introduction

The ability to make specific labelled nucleic acid probes has been central to the rapid expansion of molecular biology. The methods are part of the basic repertoire of a molecular biology laboratory. Rather than attempting to detail all of the methods historically used to label nucleic acids, we describe easy, rapid, and reproducible protocols for routine work. In the following chapter we will describe a set of frequently used hybridization protocols that utilize probes labelled as we detail here.

2. Primer extension from random oligonucleotides

Labelling DNA by random priming (1) involves annealing short oligonucleotides to denatured DNA and using these annealed oligos as primers for a DNA polymerase. The polymerase is supplied with nucleotides, one of which is labelled, and so the label is incorporated into the newly synthesized strand. The reaction conditions are chosen to produce labelled fragments with an average length of a few hundred bases, as these are suitable for most applications. This method has now replaced the older nick translation method (2) as the standard method for labelling DNA. DNA labelled by random priming may be used for Southern hybridization (see Chapter 5, Section 2.1.1) or nuclease S1 mapping (see Chapter 5, Section 5.1).

2.1 Oligonucleotides

The oligonucleotides used are typically from six to nine bases in length and of mixed, random sequence. As obtained from most suppliers (e.g. Pharmacia 27–2166–01) they are prepared from calf thymus DNA, and so each possible oligonucleotide is not represented at equal frequency. This is not normally a problem, and indeed this oligo distribution may more closely represent some target DNAs for labelling than would an even distribution. Synthesized oligonucleotides, which should have an even distribution of the possible sequences, are also available (e.g. IBI kit IB77800).

2.2 Polymerase

For labelling from a DNA template, the polymerase used is the Klenow fragment of *E. coli* DNA polymerase I. This fragment lacks the 5' to 3' exonuclease activity of the holoenzyme and so the labelled stand is synthesized only by primer extension and not by nick translation. The reaction is performed at pH 6.6 to inhibit the 3'–5' exonuclease activity of the Klenow enzyme, so that the primers and the labelled product are not degraded. It is also possible to use a variation of this protocol to label from an RNA template, using an RNA-dependent DNA polymerase. This reverse transcriptase reaction is also sometimes used to generate fragments of cDNA for constructing cDNA libraries.

2.3 Label

This method can be used to incorporate a wide-range of radioactive and non-radioactive labels into the synthesized DNA. *Protocol 1* assumes that $[\alpha\text{-}^{32}P]dCTP$ is the labelled nucleotide. If another labelled nucleotide is used then make up the unlabelled nucleotide stock with the other three nucleotides, e.g. if using $[\alpha\text{-}^{32}P]dATP$, use 500 mM of each of dCTP, dGTP, and dTTP in the stock. Some nucleotide analogues with non-radioactive labels are incorporated less efficiently than the unmodified nucleotide. In such a case the concentration of the labelled nucleotide should be increased.

2.4 Template

The normal template is linear double-stranded DNA. Closed circular DNA can also be used; the incorporation is not normally quite as good, but still adequate. It is often advisable to purify the DNA of interest away from vector sequences as this will reduce the background. DNA in low gelling temperature (l.g.t.) agarose can be used directly as indicated in *Protocol 1*. Alternatively any protocol for extracting DNA from agarose should give DNA of high enough quality. It is also possible to label single-stranded DNA, which gives a strand-specific probe. In this case there is no need to denature the DNA by boiling before carrying out the labelling reaction (*Protocol 1*, steps 2 and 3).

Protocol 1. Primer extension from random primers

Reagents

- Unlabelled nucleotide stock; 1.1 M Tris–HCl pH 8.0, 120 mM $MgCl_2$, 250 mM 2-mercaptoethanol, 500 μM dATP, 500 μM dGTP, 500 μM dTTP (store at − 20 °C)
- 2 M Hepes, pH 6.6
- Hexanucleotides (90 OD_{206} units/ml in TE)
- Klenow fragment of DNA polymerase I
- $[\alpha\text{-}^{32}P]dCTP$ (110 TBq/mmol, 370 MBq/ml), e.g. Amersham PB 10205
- TE: 10 mM Tris–HCl pH 8.0, 1 mM EDTA

Method

1. Make 5 × buffer (5 × OLB − dCTP) by mixing:
   - Hepes 2 M (pH 6.6)   125 µl
   - hexanucleotides (90 OD_{260} units/ml in TE)   75 µl
   - unlabelled nucleotide stock   50 µl

   Store in small aliquots at −20 °C.

2. Denature the DNA solution in a boiling water-bath for 3 min.[a] Use 50–250 ng DNA. Less DNA usually gives reduced incorporation, but still results in good specific activity. More DNA will reduce the specific activity of the probe.

3. Immediately place the DNA solution on ice or dry ice/ethanol.

4. Add the following to the DNA solution on ice. Do not vortex.
   - water to a final volume of   25 µl
   - 5 × OLB − dCTP   5 µl
   - Klenow enzyme   1 U
   - $[\alpha\text{-}^{32}P]dCTP$ (110 TBq/mmol, 370 MBq/ml)   1–5 µl[b]

   In order to minimize radioactive contamination of pipette tips, add the radiolabel last. For non-radioactive labels, where this is not a consideration, add the Klenow enzyme last as this starts the reaction.

5. Incubate at 37 °C for at least 1 h and at most overnight. A typical incubation would be for 4–8 h.

6. If required, purify the DNA from unincorporated nucleotides by centrifugation through a G-50 column (*Protocol 2*).

7. Store at −20 °C until use. Storage time is limited by decay of the radiolabel. Remember to denature the probe before hybridization, e.g. by boiling.

---

[a] For DNA in low gelling temperature agarose replace steps **2** and **3** with:
- Add 3 vol. of water to the gel slice and boil 7 min
- Cool rapidly to room temperature without allowing the gel to set

[b] This corresponds to 10–50 µCi.

2.5 Checking and purifying the probe

Under normal circumstances, random priming is an extremely reliable process giving greater than 50% and normally 70–95% incorporation of the label into the synthesized DNA. If necessary the incorporation can be assayed by spotting a known volume of the reaction mixture on to a glass fibre filter and washing with 5% trichloroacetic acid (TCA) to precipitate DNA, followed by an ethanol wash, and scintillation or Cerenkov counting. The amount of label in the precipitated material is compared with the total label in the reaction. In practice it is usual to proceed directly to the use of a G-50 spin column (see below).

The background in a hybridization experiment can generally be reduced if the unincorporated label is removed from the probe. However, for many routine purposes the relative intensities of signal and background are so different that purification of the probe is unnecessary. Spin column chromatography using Sephadex G-50 is the most commonly used method to purify radioactive probes.

Protocol 2. Removing unincorporated label using G-50 spun columns

Equipment and reagents

- Sephadex G-50 resin (Sigma G-50-150)
- TE: 10 mM Tris–HCl, 1 mM EDTA, pH 8
- 1 ml disposable syringe
- Polymer wool or glass wool
- Pasteur pipette
- Centrifuge capable of 1500–2000 g and 15–50 ml disposable plastic tubes to match
- Hand-held radiation monitor

Method

1. Equilibrate dry Sephadex G-50 beads (Sigma G-50-150) with TE pH 8.0 by adding beads to TE and incubating at 4 °C overnight or 65 °C for 1–2 h. Alternatively, autoclave for 15 min. Use 200 ml of TE for each 10 g of dry beads.

2. Remove and discard the plunger from a 1 ml disposable syringe.

3. Plug the bottom of the syringe barrel with a little polymer wool or glass wool. Pack the wool down into a plug with the fine end of a Pasteur pipette. The volume of the plug should then be 50–100 μl.

4. Using a Pasteur pipette, carefully fill the syringe with G-50 slurry. The volume of settled resin should be about 1 ml.

5. Cut a hole in the lid of a suitable 15 ml or 50 ml disposable plastic tube, and insert the syringe. Centrifuge at 1500–2000 g for 5 min. The resin will have compacted and partially dried. Discard the column flow-through.

6. Apply the DNA to the column in a volume of 20–100 μl. Place a microcentrifuge tube within the plastic tube beneath the syringe to collect the purified probe (see *Figure 1*).

7. Centrifuge again using the same conditions as in step **5**. If the probe was labelled with ^{32}P, then the percentage incorporation can be estimated by comparing the activity of the probe in the microcentrifuge tube with the unincorporated label on the column using a hand-held Geiger counter.

8. When radiolabelling DNA, carefully dispose of the column, which will retain 5–50% of the radioisotope used in the labelling reaction.

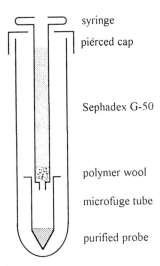

syringe

piérced cap

Sephadex G-50

polymer wool

microfuge tube

purified probe

Figure 1. Assembling a G-50 spun column.

Using the conditions of *Protocol 2*, DNA fragments of about 40–50 bp or less will be retained on the column, while larger molecules will pass through. This method can also be used to purify DNA prior to subsequent manipulations by removing salts and other small molecules. For this purpose, wash the column between steps 4 and 5 with four to five column volumes of TE.

3. RNA probes

In vitro synthesis of RNA in the presence of labelled nucleotides generates labelled RNA molecules for use as a probe. The transcription system described uses the very efficient promoters and RNA polymerases of the bacteriophages SP6, T3, and T7 (3–6). Compared with random priming, this method has a number of advantages and disadvantages. Probes made using these bacterio-phage RNA polymerases have the additional advantages of being suitable for specific applications since:

- RNA probes are strand specific
- RNA probes are normally of defined length

Strand-specific probes are useful to determine transcript orientation. They are also the best probes for some more advanced analysis such as RNase mapping (see Chapter 5, Section 5.2).

RNA probes are often used to label the ends of a large fragment of cloned DNA, such as a cosmid or lambda clone. Many such vectors have promoters for the bacteriophage RNA polymerases at the ends of the insert so that terminal probes can be made by digestion of the DNA with a restriction

125

enzyme that cuts the DNA frequently, followed by transcription from the digested DNA. Only the terminal fragment has the promoter and so this is the only one transcribed. Such terminal probes are very useful in chromosome 'walking'. They are also useful when mapping one piece of DNA on to another, for example localizing cDNAs on to cloned genomic sequences by Southern hybridization. In the presence of limiting concentrations of labelled nucleotide, most transcripts will be two kilobases or less. Although this is full-length for most templates, it means that it is not normally necessary to digest cosmid or bacteriophage lambda clones when making radiolabelled terminal probes for walking. However, such transcripts are generally too long for whole-mount RNA *in situ* procedures where the probe penetration into the sample is limited by the length of the probe. In this case it may be necessary to reduce the average size of the labelled molecules by partial alkaline hydrolysis. Overall, RNA probes are at least as good as DNA ones for most purposes and are in routine use in many laboratories.

3.1 Advantages

(a) Labelled RNA is synthesized very efficiently, with many labelled transcripts being produced from each molecule of template DNA.

(b) The probe does not need denaturation before use.

(c) RNA probes anneal better to their targets as RNA–DNA hybrids are more stable than DNA–DNA hybrids, and RNA–RNA hybrids are more stable than DNA–RNA hybrids.

(d) RNA probes are preferred in some methods for analysing transcripts as RNase A can be used to distinguish between single- and double-stranded RNA, whereas nuclease S1 is needed to distinguish between single-stranded DNA and DNA–RNA hybrids (see Chapter 5, Section 5.1).

3.2 Disadvantages

(a) The sequences to be labelled must be cloned in a vector having an appropriately positioned promoter for the bacteriophage RNA polymerase.

(b) High background is occasionally observed with some AT-rich templates.

Protocol 3. Making an RNA probe

Reagents

- 10 × transcription buffer: 400 mM Tris–HCl pH 8.0, 80 mM MgCl$_2$, 500 mM NaCl, 20 mM spermidine–HCl (store at −20 °C)
- 10 mM rATP
- 10 mM rCTP

- 10 mM rGTP
- RNase inhibitor (e.g. RNase Block II, Stratagene Cat. No. 300153/4)
- RNase-free water (diethylpyrocarbonate (DEPC) treated)

- 1 M dithiothreitol (DTT)
- RNA polymerase from SP6, T3, or T7 as appropriate
- [α-^{32}P]UTP (30 TBq/mmol, 740 MBq/ml), e.g. Amersham PB20383)

- Phenol/chloroform: phenol/chloroform/isoamyl alcohol (25:24:1) buffer saturated with 50 mM Tris–HCl pH 8
- TE: 10 mM Tris–HCl pH 8, 1 mM EDTA
- 3 M sodium acetate pH 5.2

Method

1. Linearize the plasmid DNA for template by digesting to completion with a restriction endonuclease that cuts at the distal end of the DNA of interest relative to the promoter (see *Figure 2*).

2. Remove any contaminating RNase by phenol/chloroform extraction. Mix the DNA solution with an equal volume of phenol/chloroform. Vortex and centrifuge for 3–4 min to separate the phases. Remove the upper aqueous phase to a new microcentrifuge tube.

3. Add 0.1 vol. of 3 M sodium acetate and 3 vol. of ice-cold absolute ethanol. Vortex briefly and spin in a microcentrifuge at maximum speed for 20 min to precipitate the DNA. Wash the pellet by re-centrifugation for 2–3 min with 70% ethanol and resuspend in RNase-free TE or water at 0.5–1.0 mg/ml.

4. Mix the following reagents in a microcentrifuge tube at room temperature:

- template DNA   1 μg
- 10 × transcription buffer   2.5 μl
- rATP (10 mM)   1 μl
- rCTP (10 mM)   1 μl
- rGTP (10 mM)   1 μl
- dithiothreitol (DTT, 1 M)   1 μl
- RNase inhibitor (e.g. RNase Block II) 1 U
- RNase-free water   to 25 μl
- RNA polymerase   10 U
- [α-^{32}P]UTP (30 TBq/mmol, 740 MBq/ml) 0.5–3 μl[a]

Add the radiolabel last to minimize radioactive contamination.

5. Incubate at 37°C for 30–60 min.

6. (Optional, *see text*.) To remove the template DNA, add 1 μg of RNase-free pancreatic DNase I (Sigma) and incubate for 15 min at 37°C.

7. (Optional.) Purify the probe by ethanol precipitation or G-50 spun column chromatography.

8. Store at −20°C. Storage time is limited by decay of the radiolabel.

[a] This corresponds to 20–60 μCi.

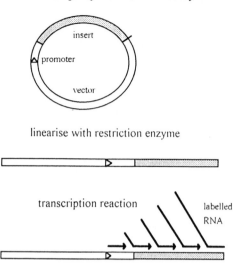

linearise with restriction enzyme

Figure 2. Making an RNA probe.

3.3 Polymerase

The three RNA polymerases used to make probes are the SP6, T3, and T7 DNA-dependent RNA polymerases (3–6). These bacteriophage polymerases share many biochemical properties and will all work using the conditions given in *Protocol 3*. They are each highly specific for their promoter sequences. Each of these polymerases are commercially available. The polymerases have a limited stability at 37 °C and so more probe can often be made by adding more polymerase after step 6 and repeating the 37 °C incubation.

3.4 Template

The template for a transcription reaction is double-stranded linear DNA. This can be readily made by digesting plasmid DNA with an appropriate restriction enzyme, as in *Figure 2*. Do not use enzymes which give 3′ protruding ends (e.g. *Pst*I) as some transcription of the complementary strand may occur.

3.5 Other labels

The bacteriophage RNA polymerases are all capable of incorporating a wide-range of nucleotide analogues into the synthesized RNA. This means that non-radioactive labels can be incorporated into probes by this method. Indeed, since higher concentrations of non-radioactive precursors can often be used, the concentration of the labelled nucleotide, which may be limiting in the reaction, can be increased, and a much greater molar amount of probe synthesized. This also allows much longer transcripts to be made. Labelled nucleotides other than UTP may be used. In this case add the three non-labelled nucleotides to the transcription reaction (*Protocol 3*, step 5).

3.6 Purifying the probe

As for probes made by random priming, the unincorporated nucleotides can be removed using a G-50 spun column (see above). It is also possible specifically to remove the template DNA by incubating the reaction with RNase-free DNase I. For most purposes, neither of these steps are necessary. However, they are essential if the probe is to be used in RNase mapping experiments.

3.7 RNase precautions

RNase is a very stable and widespread enzyme. Since the product of the transcription reaction is sensitive to this enzyme, it is sensible to minimize the risks of RNase contamination. In practice:

● wear gloves

● use diethylpyrocarbonate (DEPC) treated water (add DEPC to 0.1% and autoclave)

● keep reagents for RNA work separately

● remove RNase from the template by phenol/chloroform extraction

Using an RNase inhibitor in the transcription reaction ensures that RNase contamination is rarely a problem. RNase Block II (Stratagene Cat. No. 300153/4) works well. Alternatively, use 10 U of pancreatic RNase inhibitor. Do not use vanadyl-ribonucleoside complexes in *Protocol 3* as this RNase inhibitor will prevent RNA polymerase activity.

4. End-labelling

The use of labelled oligonucleotides for screening for sequences with a conserved sequence motif has been largely replaced by PCR (see Chapter 5, Section 3). Nevertheless, this technique is still used. Moreover, the end-labelling of oligonucleotides is also important for a number of other techniques such as primer extension (Chapter 5, Section 5.3), gel retardation (Chapter 5, Section 5.2.1), and in screening expression libraries for transcription factor binding sites (7). End-labelled restriction fragments are most commonly used as probes in footprinting experiments (Chapter 5, Section 6.2.3).

4.1 Oligonucleotides

Although each step in the chemical synthesis of an oligonucleotide is highly efficient, the accumulated errors in the synthesis of a long oligonucleotide result in a significant proportion of unwanted shorter sequences contaminating the final product. These can be removed by purifying the oligonucleotide by preparative polyacrylamide gel electrophoresis. A 15% denaturing polyacrylamide gel is suitable for purifying oligonucleotides between 9 and 50 nucleotides in length. A 0.5–1 mm thick gel is cast in the same way as in *Protocol 8*. We normally use a comb which generates a single 15 cm long slot with smaller 0.5 cm slots either side in which bromophenol blue and xylene cyanol dyes may be loaded as markers. Around 50 OD_{260} units of oligonucleotide may be loaded on to a single 1 mm gel. The concentration of the crude oligonucleotide is determined and a known amount is then lyophilized. The pellet is taken up in approximately 150 μl of deionized formamide which is then loaded on to the gel after heating to 90°C for 10 minutes. The denaturing gel is pre-electrophoresed and run as for sequencing gels (see Chapter 7). After electrophoresis the gel is transferred to a piece of Saran-Wrap or cling-film. The oligonucleotide may be visualized using a hand-held long wavelength ultraviolet lamp when the gel is placed on top of a fluorescent thin-layer chromatographic plate or alternatively an intensifying screen from an autoradiography cassette. The DNA absorbs the ultraviolet light and creates a 'shadow' on the fluorescent plate. The strong band corresponding to the full-length oligonucleotide may be cut out and placed in a 15 ml polypropylene tube. Add 2 ml of Maxam and Gilbert elution buffer (see *Protocol 8*) and place the tube on a rotary shaker at room temperature overnight. The eluted oligonucleotide may then be ethanol precipitated. Fortunately, this purification is not normally required prior to using the oligonucleotide as a probe.

Oligonucleotides are end-labelled using T4 polynucleotide kinase, which adds a phosphate group to the 5′ end. It is therefore essential that the 5′ end of the oligonucleotide is available for phosphorylation, i.e. not blocked or already phosphorylated. Oligonucleotides are synthesized without a phosphate at the 5′ end.

The oligonucleotide concentration in millimoles can be estimated by dividing the OD_{260} by:

$$(15.4 \times A) + (7.3 \times C) + (11.7 \times G) + (8.8 \times T)$$

where A, C, G, and T are the numbers of each nucleotide in the oligonucleotide. For a rough estimate, the concentration in millimoles is approximately OD_{260} divided by ten times the length in nucleotides.

End-labelled oligonucleotide probes containing the putative protein binding site are often ideal for use in gel retardation experiments (see Chapter 5, Section 6.2.1). Typically 30–50mer double-stranded oligonucleotide probes are used. The oligonucleotide is end-labelled as in *Protocol 4*. Polynucleotide kinase does not phosphorylate blunt-ends very efficiently and therefore one strand of the oligonucleotide should be labelled and then annealed to an excess of the second (unlabelled) complementary strand. The annealing reaction can be performed by mixing the complementary strands, heating to 70 °C and then cooling to room temperature. The double-stranded oligonucleotide probe can be separated from single-stranded oligonucleotides and any unincorporated label on a non-denaturing polyacrylamide gel (see *Protocol 8*).

Protocol 4. End-labelling oligonucleotides

Reagents

- 10 × T4 polynucleotide kinase (PNK) buffer: 500 mM Tris–HCl pH 8.0, 100 mM MgCl$_2$, 100 mM DTT (store at −20 °C in small aliquots)
- [γ-^{32}P]ATP (e.g. Amersham PB10218, 185 TBq/mmol, 370 MBq/ml)

Method

1. Set up the labelling reaction by mixing:
   - 10 × T4 PNK buffer   2.5 μl
   - oligonucleotides   10 pmoles
   - [γ-^{32}P]ATP   10 pmoles[a]
   - water   to 25 μl

2. Incubate at 37 °C for 30 min.

3. Store at −20 °C. Storage time is limited by decay of the label.

[a] This is 5 μl (50 μCi) of Amersham PB10218.

4.2 Specific activity

Protocol 4 uses equimolar amounts of oligonucleotide and of label. Under these conditions, approximately half of the label is transferred to the oligonucleotide. This gives a specific activity for the labelled oligonucleotide of half that of the original label. The reaction can be modified to give greater percentage incorporation of the label, but lower specific activity of the oligonucleotide, by increasing the amount of oligonucleotide five- to tenfold. This gives approximately 90% incorporation. Conversely, a threefold increase in label concentration relative to the oligo gives near 100% labelling of the oligo, though this is only about 30% incorporation of the label.

4.3 End-labelling of restriction fragments

The probes used for footprinting and gel retardation experiments are linear, double-stranded DNA molecules with a labelled group at one end (see Chapter 5, Section 6). As discussed above, oligonucleotides are the probes of choice for use in gel retardation experiments. If this is not practical a short restriction fragment may be used. Ideally this should be less than 100 bp in length to minimize the binding of non-specific proteins. Probes used in foot-printing experiments are normally somewhat longer than those used in band-shifting assays. The putative protein binding site should be within 170 bp of the labelled end to allow good resolution on sequencing gels. The site of interest should not be too near (within 50 bp) to either the 5' or 3' end of the probe otherwise interference from non-specific 'end-binding' proteins may become problematic (see Chapter 5, Section 6.2.3).

The strategy for end-labelling of DNA fragments is summarized in *Figure 3*. The DNA fragment containing the protein binding site is cloned into a plasmid vector. (Cloned plasmid DNA consistently yields better quality probes than M13 DNA clones.) The plasmid is then cut at a suitable restriction endo-nuclease cleavage site. The 5' or 3' end is then labelled with ^{32}P. The DNA is

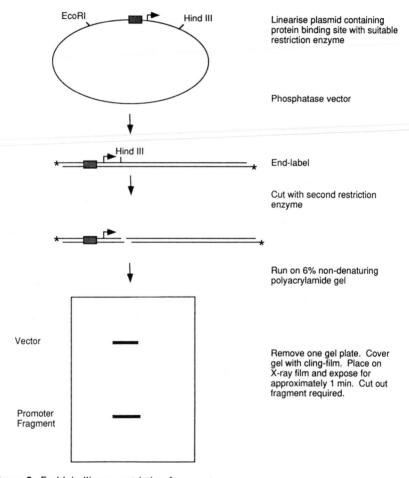

Figure 3. End-labelling a restriction fragment.

then re-digested with a second restriction endonuclease to provide a probe that is labelled at one end only. *Protocols 5–7* outline the procedures used for 5' or 3' end-labelling of DNA. Probes labelled at the 5' and 3' ends are useful for looking at the contacts a protein makes with both strands of a DNA molecule in footprinting experiments. Finally the desired probe is separated from any other labelled fragments by non-denaturing polyacrylamide gel electrophoresis as detailed in *Protocol 8*.

Protocol 5. Preparation of 5' end-labelled probes

Reagents

- Calf intestinal alkaline phosphatase (e.g. Boehringer-Mannheim 713023)
- Phenol (buffer-saturated with 50 mM Tris–HCl, pH 8)
- 3 M sodium acetate pH 5.2
- 10 × PNK buffer and radiolabelled nucleotide as in *Protocol 4*
- Appropriate restriction endonucleases and buffers (supplied with the enzyme)
- 10 × TBE buffer pH 8.3: 108 g Tris base, 55 g boric acid, 9.3 g EDTA (Na$^+$ salt) made up to 1 litre
- Native sample loading buffer: 40% glycerol, 2.5 × TBE, 50 mM EDTA, 0.1% bromophenol blue

Method

1. 10 μg of plasmid is digested with the first restriction endonuclease (which should leave 5' overhangs) in 50 μl of the appropriate restriction endonuclease digestion buffer.

2. After 1 h dilute the reaction to 100 μl with water and add 5 U of calf intestinal phosphatase. Incubate at 37 °C for a further 30 min.

3. Heat to 70 °C for 10 min to inactivate the enzymes, then add an equal volume of buffer-saturated phenol. Mix by vortexing and then centrifuge for 3 min to separate the phases. Transfer the aqueous (upper) phase to a new microcentrifuge tube.

4. Add 10 μl of 3 M sodium acetate pH 5.2, and 250 μl of ethanol. Mix by vortexing briefly, and pellet by centrifugation in a microcentrifuge for 25 min. Carefully decant the supernatant and wash the pellet in 70% ethanol. Dry briefly.

5. Re-dissolve the DNA in 12 μl of water.

6. Use 3 μl of DNA for each labelling reaction (i.e. 2.5 μg of linearized plasmid). Set-up a labelling reaction as follows (PNK buffer and radio-labelled nucleotide are as described in *Protocol 4*):
   - 3 μl of DNA
   - 1 μl of 10 × PNK buffer
   - 5 μl of [γ-^{32}P]ATP
   - 1 μl of water

Protocol 5. *Continued*

7. Incubate at 37 °C for 30 min. Inactivate the enzyme by heating to 65 °C for 10 min.

8. Dilute the end-labelled DNA to 40 μl by adding:[a]
   - 4 μl of the appropriate restriction endonuclease buffer
   - 1 μl of the restriction endonuclease for the second digest (usually 8–12 U)
   - 25 μl of water

9. Incubate at 37 °C for at least 1 h. Add 10 μl of native sample loading buffer. Load on a 6% non-denaturing polyacrylamide gel as described in *Protocol 8*.

---

[a] Should the second restriction endonuclease not cut, steps **3** and **4** should be repeated to remove proteins and precipitate the DNA before carrying out the digestion in a final volume of 40 μl. The method outlined here normally avoids the need for phenol extraction and precipitation of the DNA.

Protocol 6. Labelling a 3′ recessed end[a]

Reagents

- Phenol (buffer-saturated with 50 mM Tris–HCl pH 8)
- 3 M sodium acetate pH 5.2
- Appropriate restriction endonucleases and buffers
- Native sample loading buffer: 40% glycerol, 2.5 × TBE (*Protocol 8*), 50 mM EDTA, 0.1% bromophenol blue
- 10 × Klenow buffer: 500 mM Tris–HCl pH 7.6, 100 mM MgCl$_2$

- [α-^{32}P]dXTP, where X is a nucleotide complementary to one found in the 5′ overhang (e.g. [α-^{32}P]dCTP as in *Protocol 1*)
- 1 mM dNTP − dXTP mix (containing the three dNTPs other than dXTP, or as many as are required to fill in the overhang)
- Klenow fragment of *E. coli* DNA polymerase I

Method

1. Follow *Protocol 5*, step 1 and steps 3–5, step 2 is not necessary.

2. Set up a labelling reaction as follows:
   - 3 μl DNA (2.5 μg)
   - 1.1 μl of 10 × Klenow buffer
   - 5 μl of [α-^{32}P]dXTP
   - 1 μl 1 mM dNTP −dXTP mix
   - 1 μl (5 U) Klenow enzyme.[a]

3. Incubate at room temperature for 15 min. Inactivate the enzyme by heating to 65 °C for 10 min.

4. Follow *Protocol 5*, steps 8 and 9.

---

[a] T7 DNA polymerase or the reverse transcriptase enzymes may also be used to efficiently label 5′ overhangs.

Protocol 7. Labelling a 3′ overhang end

Reagents
- Phenol (buffer-saturated with 50 mM Tris–HCl)
- 3 M sodium acetate pH 5.2
- Appropriate restriction endonucleases and buffers
- 10 × terminal transferase buffer: 1 M potassium cacodylate pH 7.2, 20 mM CaCl$_2$, 2 mM DTT
- [α-^{32}P]dATP (Amersham PB10204: 110 TBq/mmol, 370 MBq/ml)
- Terminal deoxynucleotidyl transferase
- Native sample loading buffer: 40% glycerol, 2.5 × TBE, 50 mM EDTA, 0.1% bromophenol blue

Method
1. Follow *Protocol 5*, step 1 and steps 3–5.
2. Set up a labelling reaction as follows:
   - 3 μl DNA (2.5 μg)
   - 4 μl of 10 × terminal transferase buffer
   - 5 μl of [α-^{32}P]dATP
   - 1 μl (10 U) of terminal deoxynucleotidyl transferase
   - water to a final volume of 40 μl
3. Incubate at 37°C for 60 min.
4. Follow *Protocol 5*, steps 3–5.
5. Perform the second restriction endonuclease digestion in a final volume of 40 μl in the appropriate restriction enzyme buffer.
6. As *Protocol 5*, step 9.

4.4 Purification of the probe

If an oligonucleotide needs to be purified, this should be done before labelling as indicated in Section 4.1. Residual [γ-^{32}P]ATP will not normally affect subsequent use of a labelled oligonucleotide. If necessary, the unincorporated label can be removed by ethanol precipitation, if the oligo is long enough to be precipitated (15–20 bases), or by gel electrophoresis (see Section 4.1), or preparative chromatography. Preparative chromatography may be carried out by filling a disposable 1 ml pipette with Sephadex G-50 (equilibrated in TE). The bottom of the pipette may be blocked with a glass bead or glass wool to prevent leakage of the column matrix. The labelled oligonucleotide is then applied to the top of the column formed in the pipette and allowed to run through the gel under gravity flow. The labelled oligonucleotide will be eluted first and the unincorporated label will run more slowly down the column. The progress of the oligonucleotide down the column may be monitored using a hand-held monitor and the oligonucleotide is collected in a few drops from the column. If an oligonucleotide probe is to be used in gel retardation or methylation interference experiments the double-stranded probe is best purified by gel electrophoresis as described in *Protocol 8*. This procedure can easily separate single-stranded oligonucleotides from double-stranded oligonucleotide. End-labelled DNA fragments for use as probes are separated from other labelled species as described in *Protocol 8*.

Protocol 8. Polyacrylamide gel electrophoresis

Equipment and reagents

- Polyacrylamide gel apparatus—the volumes of the buffers utilized in making the gel should be adjusted as appropriate for the size of apparatus available
- Scalpel
- Hand-held radiation monitor
- Speed-Vac desiccator
- Scintillation counter
- X-ray film (e.g. Kodak XAR-5)
- Large (19 gauge) syringe needle
- Siliconized glass wool
- 30% acrylamide stock solution: 29% acryla-
- mide, 1% bisacrylamide which has been filtered (through Whatman filter paper) and stored at 4°C
- 10 × TBE buffer (*Protocol 5*)
- 10% ammonium persulfate
- *N,N,N',N'*-Tetramethylethylenediamine (TEMED), e.g. BDH Cat. No. 44308
- Maxam and Gilbert elution buffer (8): 500 mM ammonium acetate, 10 mM magnesium acetate, 1 mM EDTA, pH 8.0, 0.1% SDS

Method

1. Prepare a non-denaturing 6% polyacrylamide gel by mixing:
   - 30 ml of 30% acrylamide stock solution
   - 7.5 ml of 10 × TBE buffer
   - 112.5 ml water.

2. Add 500 μl of 10% ammonium persulfate and 50 μl of TEMED.

3. Pour the gel. We generally use two 20 × 20 cm gel plates, one of which has been cut to form 'ears' to allow the insertion of a comb. The plates are separated by 1 mm spacers and taped around the edges. For probe purification use a comb which creates slots around 12 mm wide. Allow the gel to polymerize for around 1 h.

4. Set-up the gel with 0.5 × TBE in the buffer reservoirs. Pre-electrophoresis of non-denaturing gels is not usually required. Load the samples to be separated. Run at 20 V/cm until the bromophenol blue dye reaches the bottom of the gel (or as long as is necessary in order to separate the probe from vector DNA fragments).

5. Remove the gel from the apparatus and carefully remove one gel plate. Cover the gel with cling-film.

6. In a dark-room place the gel on X-ray film (e.g. Kodak XAR-5) and expose for 1 min. Draw around the edges of the gel plate with a pencil to mark the position of the gel on the film.

7. Develop the film. Locate the radioactive band of interest and excise with a scalpel (taking the minimum amount of acrylamide). Check with a Geiger counter that most of the counts have been excised from the gel.

8. Use a large (19 gauge) syringe needle to puncture a hole in the bottom of a lidless microcentrifuge tube. Place the sliver of acrylamide into this tube. Place this tube in an intact microcentrifuge tube. Spin in a microcentrifuge until all the acrylamide is in the lower tube. This crushes the acrylamide.[a]

9. Add 300 μl of Maxam and Gilbert elution buffer to the upper micro-centrifuge tube. Centrifuge again briefly.

10. Discard the upper tube and cap the lower tube. Mix by vortexing and leave on a rotary shaker overnight at room temperature. Centrifuge for 3 min and retain the supernatant.

11. Prepare another pair of microcentrifuge tubes, one with a pierced bottom, as above. Pack the bottom third of the upper microcentrifuge tube with siliconized glass wool. Transfer the supernatant from step **10** to the upper microcentrifuge tube and spin briefly in a micro-centrifuge. Discard the upper microcentrifuge tube.[b]

12. Add 2.5 vol. of ethanol to the eluted probe. Mix by vortexing, and centrifuge for 30 min in a microcentrifuge.

13. Discard the supernatant carefully. Wash the pellet with 70% ethanol. Dry the pellet in a Speed-Vac desiccator and take it up in 30 μl of water. Centrifuge for 1 min and transfer supernatant to a new tube. Count an aliquot in a scintillation counter (Cerenkov). If the probe is to be used in footprinting or gel retardation experiments adjust the volume to give 10^4 c.p.m./μl.

[a] If eluting oligonucleotides from the gel there is no need to crush the acrylamide.
[b] Elutip D-columns (Schleicher and Schuell) may also be used to purify the probe DNA.

4.5 Maxam and Gilbert sequencing reactions to provide marker tracks

The Maxam and Gilbert sequencing method requires that the target DNA be labelled at only one end. Probes prepared as in *Protocols 5–7* are useful substrates for performing Maxam and Gilbert sequencing reactions. Maxam and Gilbert have written detailed protocols for performing the reactions involved in chemical sequencing on which the methylation interference protocol described in Chapter 5 is based (9). When running footprinting and methylation interference gels it is normal to load a marker lane alongside the samples loaded. *Protocol 9* describes how to perform the Maxam and Gilbert G + A reactions on an end-labelled probe. This is the simplest of the chemical cleavage reactions to perform and is therefore most commonly used as a marker lane. A Savant Speed-Vac desiccator or equivalent is needed to perform these reactions efficiently as all traces of piperidine must be removed after the cleavage reaction, otherwise smeary bands will result.

Protocol 9. Maxam and Gilbert G + A sequence reaction

Equipment and reagents

- Speed-Vac desiccator
- Salmon sperm DNA
- Freshly diluted 2% formic acid
- 1 M piperidine (=10%)

- Formamide loading dyes: 95% formamide, 0.05% bromophenol blue, 0.05% xylene cyanol FF

Method

1. Prepare and purify the probe as described above (*Protocols 5–8*).
2. Dry down 5×10^4 c.p.m. (Cerenkov) of probe with 1 μg of salmon sperm DNA in a Speed-Vac desiccator.
3. Add 2.5 μl of freshly diluted 2% formic acid.
4. Incubate for 10 min at 37°C. Quench in a dry ice/ethanol bath.
5. Lyophilize in a Speed-Vac for at least 30 min.
6. Take up in 100 μl of 1 M (= 10%) piperidine.
7. Incubate for 30 min at 90°C in a tightly capped microcentrifuge tube.
8. Quench the reaction on dry ice/ethanol. Lyophilize for at least 90 min until all visible traces of piperidine have been removed.
9. Take up in 40 μl of water and lyophilize for at least 45 min.
10. Take up in 10 μl of formamide loading dyes.
11. Heat for 5 min at 90°C to denature before loading on a sequencing gel as a marker lane.

---

5. Non-radioactive probes

Although probes have historically been made by incorporating a radioisotope into a nucleic acid, allowing detection by autoradiography, there is now available a wide-range of non-radioactive labels that can be used to label nucleic acids. Compared to radioactive labelling, this has several advantages in:

- safety
- stability of probe
- efficiency of the labelling reaction
- detection *in situ*
- time taken to detect signal.

Non-radioactive labels are normally incorporated into nucleic acid probes by using a non-radioactive nucleotide derivative in place of the radiolabelled nucleotide in the random priming or RNA transcription reactions described above. Commonly used nucleotide derivatives used for this purpose include digoxygenin-UTP and -dUTP, and biotin-dUTP (supplied by Boehringer-Mannheim). After hybridization, these are detected by an antibody or avidin respectively, followed by a colour or chemiluminescent reaction catalysed by alkaline phosphatase or peroxidase linked to the antibody or avidin. FITC- and other fluorescent derivatives of nucleotides can also be incorporated directly into the probe, or conjugated to the antibody or avidin in place of an enzyme. Since the labelled nucleotide need not be at a limiting concentration, the amount and average length of the probe may be considerably greater than for radiolabelled probes. Boehringer-Mannheim supply all of the reagents and have also published a useful guide to non-radioactive detection (10).

5.1 Sensitivity

The sensitivity of non-radioactive probes is currently no better than that of radioactive ones, and usually not quite as good. The greater number of steps involved in detecting the signal after hybridization also means that more bench time is required for each experiment than with radioactive probes. However, the result is usually obtained more quickly, as colour reactions typically take hours for a faint signal whereas autoradiography takes days.

5.2 Use for *in situ* hybridization

The one area in which non-radioactive probes have a clear advantage is in *in situ* hybridization. When the probe is detected by fluorescence or colour reaction, the signal is at the exact location of the annealed probe, whereas radioactive probes can only be visualized as silver grains in a photographic emulsion some distance away from the actual annealed probe.

5.3 Specific activity

The specific activity of nucleotides with non-radioactive labels may be too high as supplied. For example, for a label such as biotin or digoxygenin which is to be detected by a large molecule, steric effects mean that there is no point in incorporating the label more often than once per 20 or so bases. The labelling reaction may also work better if the label is diluted with some unlabelled nucleotide. Since the polymerase may not incorporate labelled nucleotides as efficiently as unlabelled ones, the optimal amount of unlabelled nucleotide varies from case to case.

5.4 Oligonucleotides

Oligonucleotides can be synthesized by incorporating labelled nucleotide analogues as they are made. This has advantages in terms of labelling efficiency and ease, but can be expensive. A wide range of labelled precursors are now available.

6. PCR

Probes can be made by using the polymerase chain reaction (PCR) with labelled nucleotides. Vector-free probes may be produced by means of the polymerase chain reaction, thus avoiding the need for restriction enzyme digestion, preparative gel electrophoresis, and elution of DNA fragments from vectors. Either radiolabelled or non-radiolabelled probes (e.g. biotinyl-ated) probes can be produced. Probes which incorporate biotin-11-dUTP or digoxygenin-11-UTP are very stable and sufficient can be produced in one set of PCR cycles for several experiments (11). PCR can also be used to make probes directly from bacterial colonies avoiding the need to prepare DNA from liquid cultures (12). PCR is suited to making single-stranded DNA probes. By using a single primer and performing 20–30 reaction cycles large amounts of single-stranded probe can be produced. A PCR protocol is given in Chapter 5 (*Protocol 6*). This may be adapted for the incorporation of non-radioactive or radioactive labels in the following ways:

(a) Non-radioactive labels. The concentration of dTTP in the 5 mM dNTP mix is reduced to 3.25 mM, and 1.75 mM digoxygenin-11-dUTP (or biotin-11-dUTP) label is added to the mix. Digoxygenin-labelled DNA is synthesized 50% less efficiently than unlabelled DNA. The labelled DNA may be separated from any unincorporated dNTPs by ethanol precipita-tion. The probe DNA **must not** be phenol extracted.

(b) Radioactive labels. Reduce the concentration of the nucleotides in the dNTP mix to 2 mM and that of the dCTP (or dATP depending on the radio-label used) to 1 mM. In addition to the cold nucleotides 5 μl of [α-^{32}P]dCTP (110 TBq/mmol, 370 MBq/ml, e.g. Amersham PB 10205) is added to the PCR reaction which is performed in a final volume of 100 μl.

As discussed in Chapter 5, Section 3 the optimum conditions for any PCR reaction must be determined individually. The reaction conditions given above work well for the generation of probes. Any probe synthesized should be gel purified before use. It is usually best to amplify a DNA segment by PCR, gel purify the DNA fragment required, and then label the DNA as in *Protocol 1*.

References

1. Feinberg, A. P. and Vogelstein, B. (1983). *Anal. Biochem.*, **132**, 6.
2. Rigby, P. W. J., Dieckmann, M., Rhodes, C., and Berg, P. (1977). *J. Mol. Biol.*, **113**, 237.
3. Schenborn, E. T. and Meirendorf, R. C. (1985). *Nucleic Acids Res.*, **13**, 6223.
4. Sambrook, J., Fritch, E. F., and Maniatis, T. (ed.) (1989). *Molecular cloning: a laboratory manual* (2nd edn). Cold Spring Harbor Laboratory, Cold Spring Harbor, NY.
5. Green, M. R., Maniatis, R., and Melton, D. A. (1983). *Cell*, **32**, 681.
6. Tabor, S. and Richardson, C. C. (1985). *Proc. Natl. Acad. Sci. USA*, **82**, 1074.
7. Singh, H., Lebowitz, J. H., Baldwin, A. S., and Sharp, P. A. (1988). *Cell*, **52**, 415.
8. Maxam, A. and Gilbert, W. (1977). *Proc. Natl. Acad. Sci. USA*, **74**, 560.
9. Maxam, A. and Gilbert, W. (1980). In *Methods in enzymology*, Vol. 65, pp. 499–510. Academic Press, London.
10. *The Dig system user's guide for filter hybridisation.* (1993). Boehringer-Mannheim GmbH Biochemica, PO Box 31 01 20, D-6800 Mannheim 31, Germany.
11. Lo, Y.-M. D., Mehal, W. Z., and Fleming, K. A. (1988). *Nucleic Acids Res.*, **16**, 8719.
12. Taylor, G. R. (1991). In *PCR: a practical approach* (ed. M. J. McPherson, P. Quirke, and G. R. Taylor), p. 10. IRL Press, Oxford.

5

The utilization of cloned DNAs to study gene organization and expression

HUW D. PARRY and LUKE ALPHEY

1. Introduction

In this chapter we present a collection of techniques for the analysis of gene expression. The methods described are those most commonly employed in studies of eukaryotic gene regulation. They utilize cloned genes as the subjects of study and as a means of generating nucleic acid probes. Southern hybridization is a powerful technique for providing a physical map of a gene on the chromosome. Information on the number of copies of the gene in the genome and the degree of similarity of the gene under study to other known homologues may also be obtained. Similarly, northern blotting provides information on the size and abundance of specific mRNA molecules in preparations of poly(A)$^+$ RNA. A range of techniques are available to determine the 5′ and 3′ termini of an RNA molecule and the positions of any splice sites within them. RNase and S1 mapping provide both qualitative and quantitative information about messenger RNA (mRNA) structure. Many eukaryotic genes are regulated at the transcriptional level. Sequencing of a promoter region often reveals elements having homology to known transcription factor binding sites. The gel retardation and footprinting methods provide useful information on the binding of specific transcription factors to the promoter regions of genes. Once sequence-specific DNA binding proteins have been identified they may be isolated by DNA affinity chromatography (1). Purified proteins may then be used for *in vitro* transcription studies to build up a complete understanding of the molecular interactions driving gene expression.

2. Hybridization

The hybridization of complementary single-stranded nucleic acids forms a core technique in molecular biology, with a vast number of variants. Rather than attempt to cover all the variations, we present basic techniques for binding nucleic acids to filters and then hybridizing labelled probes to them.

2.1 Immobilizing nucleic acids on filters

In this section we describe widely used methods for transferring both DNA and RNA on to a membrane for hybridization.

2.1.1 DNA filters (Southern blotting)

Agarose gel electrophoresis is the standard method for separating DNA molecules that have been digested with one or more restriction enzymes on the basis of their size. Linear fragments of DNA up to 50 kb in size can be resolved on agarose gels. The amount of agarose in the gel determines the size range of DNA fragments that can be separated (see *Table 1*). Agarose gels are prepared by boiling agarose and buffer together until the agarose has dissolved. Tris–acetate (TAE) and Tris–borate (TBE) are the most commonly used buffers (see *Table 2*). TAE has a much lower buffering capacity than TBE. During extended electrophoresis the buffer may become exhausted. Therefore either replacement of the buffer or recirculation of buffer between anode and cathode is recommended. The fluorescent intercalating

Table 1. Range of separation of DNA fragments in gels containing different amounts of agarose

Amount of agarose in gel (% w/v)	Range of separation of linear DNA (kb)
0.3	5.0–50
0.5	1.0–25
0.8	0.7–8.0
1.0	0.5–7.0
1.2	0.4–6.0
1.5	0.2–3.0
2.0	0.1–2.0

Table 2. Composition of electrophoresis buffers

Buffer	Concentration of stock solution	Working solution
Tris–acetate (TAE)	50 ×: 242 g Tris base 57.1 ml glacial acetic acid 100 ml 500 mM EDTA pH 8.0	1 ×: 40 mM Tris–acetate 1 mM EDTA
Tris–borate (TBE)	10 ×: 108 g Tris base 55 g boric acid 40 ml 500 mM EDTA pH 8.0	1 ×:[a] 90 mM Tris–borate 1 mM EDTA

[a] 0.5 × TBE may also be used.

dye, ethidium bromide, may be added to the buffer in the electrophoresis tank and to the gel before pouring at a concentration of 0.5 µg/ml. As ethidium bromide is a powerful mutagen, to minimize the risk of contamination gels may be run in the absence of the dye and the gel stained after running in a solution of the gel buffer to which ethidium bromide has been added. The same solution can be reused to stain many gels if stored in the dark. After electrophoresis DNA can be detected by direct examination of the gel placed on an ultraviolet light box. Ethidium bromide is stored as a 10 mg/ml stock solution in a dark bottle. In our laboratory we routinely use the Horizon gel casting and gel electrophoresis systems supplied by Gibco-BRL. Agarose gel electrophoresis is described extensively in ref. 2. DNA fragments up to 10 Mb in length from entire chromosomes can be resolved by pulsed field electro-phoresis (3, 4).

Although it is possible to immobilize the DNA within agarose gels, the DNA is normally denatured and then transferred to a membrane in a process known as Southern blotting after its originator E. M. Southern (see *Figure 1*) (5). Subsequent autoradiography of the filter results in a specific pattern of bands that are homologous to the DNA or RNA probe.

Protocol 1. Transferring DNA from agarose gels to nitrocellulose (Southern blotting)

Equipment and reagents

- Rotary shaker (e.g. 'The Belly Dancer' supplied by Stovall Life Sciences Inc.)
- Nitrocellulose or other hybridization membrane (see text)
- Whatman 3 MM paper
- Blotting paper or absorbent towels
- Saran-Wrap or cling-film

- Glass plate
- Oven set to 80 °C
- Depurination solution: 0.25 M HCl
- Denaturation buffer: 1.5 M NaCl, 0.5 M NaOH
- Neutralization buffer: 1 M ammonium acetate, 20 mM NaOH

Method

1. (Optional.) Depurinate the DNA by washing the gel in 0.25 M HCl for 15 min.

2. Denature the DNA by washing the gel in denaturation buffer for 30 min with gentle agitation on a rotary shaker.

3. Neutralize the gel by rinsing in neutralization buffer and then washing in the same buffer for 30 min with gentle agitation.

4. Cut a piece of nitrocellulose[a] to the required dimensions for the part of the gel to be blotted. Also cut to this size four pieces of 3 MM paper and a stack of blotting paper or absorbent paper towels approximately 2–8 cm thick. Mark the nitrocellulose to identify the experiment and the orientation of the gel.[b]

Protocol 1. *Continued*

5. Wet the nitrocellulose and two of the pieces of 3 MM paper by floating them on neutralization solution.

6. Place the gel on a large piece of cling-film.

7. Lay the nitrocellulose on to the gel, and the wet 3 MM paper on top of the nitrocellulose. At each stage carefully remove any air bubbles by rolling a pipette over the assembly. Stack the remaining 3 MM paper and blotting paper on top, as shown in *Figure 1A*.

8. Wrap the cling-film around the gel and paper stack to reduce drying by evaporation.

9. Put a glass plate on top of the stack and weigh it down with 100–500 g, depending on gel stiffness.[c]

10. Leave 0.5–2 h.

11. Remove the nitrocellulose filter and mark the positions of the gel slots with a pencil if required.

12. Bake the filters for 1 h at 80 °C between sheets of 3 MM paper.

[a] Handle filters carefully at all times; wear gloves and hold the membrane by the edges with blunt forceps. Creases and grease spots seriously disrupt transfer and hybridization. Other hybridization membranes may also be used (see below).
[b] To obtain two copies of the gel, double these quantities and assemble them as shown in *Figure 1B*.
[c] The weight is simply to improve the contact between the layers. Too much weight will crush the gel, impairing transfer.

A generally applicable protocol is given in *Protocol 1*. The following points should be noted.

(a) Originally, the membrane used was always nitrocellulose, as described in *Protocol 1*. This is still perfectly satisfactory for many applications, but various nylon membranes (e.g. Hybond N supplied by Amersham International) are now available, which have the advantages of:

- durability
- covalent attachment of the nucleic acid
- range of permissible transfer buffers.

Nylon membranes often perform far better than nitrocellulose. The greater toughness of the membrane is a distinct advantage, especially for larger filters. The principal disadvantage of nylon membranes is the higher background levels often observed with some non-radioactive detection methods. Positively charged nylon membranes (e.g. Hybond N+) have a higher capacity for nucleic acid and so are generally preferred to unmodified nylon. Since nylon membranes tolerate alkali conditions well, the

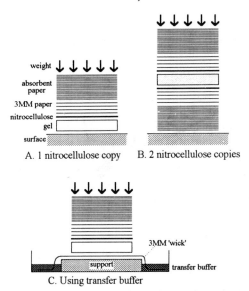

Figure 1. Transferring DNA from a gel to a nitrocellulose membrane.

transfer of DNA can be performed in denaturation buffer, omitting step 3 of *Protocol 1*. The nucleic acid can be linked to the membrane by baking in a vacuum oven at 80 °C, UV crosslinking, or other methods. Follow the manufacturers' instructions to carry out these procedures.

(b) All membranes for blotting and hybridization must be stored and handled with great care at all times. Any damage or surface contamination of the membrane can lead to poor transfer or artefactual signals. Wear gloves at all times and manipulate the filter by holding it at the edges with blunt forceps.

(c) DNA larger than about 8–10 kb is inefficiently transferred on to the membrane as it moves only slowly out of the gel. This can be remedied by partial acid depurination of DNA prior to transfer (*Protocol 1*, step 1). Denaturation then liberates shorter, more mobile fragments. However, it is important not to depurinate excessively as this will result in DNA fragments that are too short to bind efficiently to the membrane. Consequently, this treatment somewhat reduces the efficiency with which small DNAs are detected.

(d) It may be possible to increase the efficiency of transfer by adding more liquid to the bottom of the gel, rather than just relying on the liquid contained within the gel. This is the situation illustrated in *Figure 1C* where transfer buffer is drawn up through the wick of 3 MM paper, through the gel, and into the stack of absorbent paper. The transfer buffer can be neutralization buffer or 20 × sodium saline citrate (SSC), for nitrocellulose membranes. For nylon membranes, denaturation solution or 2 × SSC can also be used. Of course only one copy can be obtained by this method. Place a non-absorbent barrier around the gel, e.g. cling-film, to prevent the absorbent paper coming into contact with the wick. Any 'short-circuiting' will prevent the even transfer of the nucleic acid.

(e) Vacuum transfer is somewhat faster and more efficient than capillary transfer. There are a number of commercially available systems which work well but tend to be rather fiddly. Damage to the gel is more of a problem than using the standard method given in *Protocol 1*. The procedure for using vacuum blotting is essentially the same except that buffer is drawn through the gel by vacuum rather than capillary action. Care must be taken not to use too high a vacuum or the gel will become compressed and less porous.

(f) Nucleic acid can be transferred from a gel by electrophoresis. This requires a nylon membrane as the high ionic strength buffers needed for nitrocellulose to bind DNA have too low an electrical resistance. This method is not often used, although there are commercially available systems. It may be the best way to transfer nucleic acids from polyacrylamide gels.

2.1.2 RNA filters (northern blotting)

Transfer of RNA to filters is often called northern blotting (6) to contrast it with the transfer of DNA to filters (see above). The procedures are very similar. RNA is separated by electrophoresis through a denaturing agarose gel and then is transferred to a membrane. RNA is extremely sensitive to nuclease degradation. It is essential to take the precautions of wearing gloves, using disposable plasticware direct from the suppliers packaging and treating all glassware with a 0.1% solution of diethylpyrocarbonate (DEPC) to destroy RNase.

i. Preparation of RNA

A large number of techniques have been developed to purify RNA (2, 8). The method selected will depend on whether the starting material is tissue culture cells or a tissue sample, and whether a particular subcellular fraction is selected to enrich for a particular type of RNA. The extraction of RNA using acid phenol is one of the simplest and most reproducible techniques to employ for cells grown in tissue culture (2,8). mRNAs can be separated from other RNAs by means of oligo(dT)-cellulose chromatography.

ii. Gel electrophoresis

Either formaldehyde or glyoxal/dimethysulfoxide (DMSO) are used to denature RNA prior to agarose gel electrophoresis. Glyoxal/DMSO gels take longer to run than formaldehyde gels but the bands detected by northern hybridization are much more distinct. As glyoxal reacts with ethidium bromide the gels are cast and run in the absence of the dye. *Protocol 2* outlines gel electrophoresis after denaturation of RNA using glyoxal/DMSO.

Protocol 2. Electrophoresis of RNA after denaturation with glyoxal/DMSO

Equipment and reagents

- Gel tank with buffer chambers at either end connected with plastic tubing which passes through a peristaltic pump
- Power pack
- 6 M deionized glyoxal[a]

- Dimethylsulfoxide (DMSO)
- 0.1 M NaH_2PO_4 pH 7.0[b]
- Glyoxal loading buffer: 50% glycerol, 10 mM NaH_2PO_4 pH 7.0, 0.25% xylene cyanol, 0.25% bromophenol blue

Method

1. In a sterile RNase-free microcentrifuge tube mix:
   - DMSO   20 μl
   - 0.1 M NaH_2PO_4 pH 7.0   4 μl
   - glyoxal   5.9 μl
   - RNA (up to 10 μg) and H_2O 10.1 μl

   Incubate at 50°C for 60 min.

2. Prepare a 1% (w/v) agarose gel in 10 mM NaH_2PO_4 pH 7.0 whilst the samples are incubating.

3. Cool samples on ice and add 11 μl of glyoxal loading buffer to each sample. Immediately load the samples into the wells of the gel. Glyoxalated RNA size markers (purchased from any reputable company) should also be loaded alongside.

4. Run the gel submerged in 10 mM $NaPO_4$ at 3–5 V/cm. A peristaltic pump should be used to recirculate the buffer between the buffer chambers in order to prevent H^+ gradients forming during electrophoresis. Glyoxal dissociates from RNA at pH 8.0. If a peristaltic pump is not available the buffer may be changed by hand at regular intervals (20–30 min).

5. Stop the gel running when the bromophenol blue has migrated approximately half-way down the gel. The marker lane may be cut from the gel and stained with ethidium bromide before re-aligning the marker lane and photographing the gel under ultraviolet light. The gel should then be transferred to a nylon membrane (e.g. Hybond N supplied by Amersham) as described in the text.

[a] Deionize 6 M glyoxal by pouring over a small mixed-bed resin column (e.g. Bio-Rad AG 501-X8). Store at −20°C in small, tightly capped aliquots.
[b] Treat with DEPC and sterilize by autoclaving.

Glyoxal–RNA gels can be blotted straight away (*Protocol 1*, step 4). For formaldehyde gels, the formaldehyde should first be removed by rinsing the gel several times in DEPC treated (RNase-free) water. When blotting RNAs longer than 3 kb it is worthwhile subjecting the RNA in the gel to partial hydrolysis with 0.05 M NaOH for 15–20 minutes. This will improve the transfer efficiency of the RNA. The gel can then be blotted by the method of *Protocol 1*, steps 4 to 12. Nylon membranes are usually used for RNA blots, and capillary blotting with a liquid reservoir may give a more efficient transfer (see Section 2.1.1). When RNA has been denatured with glyoxal, remove the glyoxal by baking or washing in 20 mM Tris–HCl pH 8.0 at 65 °C or 0.2 × SSC/0.5% SDS at 90–100 °C.

2.1.3 Transferring DNA from colonies or plaques to membranes

DNA from bacterial colonies or bacteriophage plaques immobilized on membranes permits the screening of libraries for clones of interest and the selection of clones after any manipulation of recombinant DNA. The protocol refers to colonies for convenience only. The procedure is similar for bacteriophage lambda plaques for which special consideration is given in Chapter 2. M13 plaques contain single-stranded DNA and so do not need denaturing. Thus steps 1 and 6 to 9 of *Protocol 2* should be omitted when screening M13 plaques.

Protocol 3. Screening bacteria or bacteriophage DNA by transfer to membrane

Equipment and reagents

- Whatman 3 MM paper
- Nitrocellulose discs corresponding to the size of the plates to be screened
- Two pairs of blunt-ended forceps
- Syringe needle (19 gauge)
- Oven set to 80 °C
- Denaturation solution: 1.5 M NaCl, 0.5 M NaOH
- Neutralization solution: 1 M ammonium acetate, 20 mM NaOH

Method

1. Saturate a stack of two to four sheets of Whatman 3 MM paper with denaturation solution.

2. Label a nitrocellulose disc with a soft pencil or marker pen.[a]

3. Using two pairs of blunt-ended forceps, place a nitrocellulose disc on to the plate. Avoid trapping air bubbles.

4. Key the filter to the plate by stabbing through the filter and agar with a needle to make a pattern of holes.[b]

5. When the disc is completely wet, peel it off without smearing the colonies.

6. Place colony (or plaque) side up on the paper soaked in denaturation solution[c] for 2–5 min.

---

7. Dry the filters briefly by placing them colony side up on to dry 3 MM paper.

8. Saturate a stack of two to four sheets of Whatman 3 MM paper with neutralization solution.

9. Lay the filter colony side up on to the paper soaked in neutralization solution for 5 min.

10. Dry the filters briefly by placing them colony side up on to dry 3 MM paper.

11. Bake the filters for 1 h at 80 °C between sheets of 3 MM paper.

[a] The ink from some marker pens is washed off the filter during hybridization. Pencil and other marker inks can bind probe.

[b] Attaching an ink-filled syringe to the needle makes the marks more obvious, but can be messy. Use permanent ink. When taking multiple lifts from the same plate, use the same set of holes.

[c] Shallow pools of solution can be used in place of soaked paper. Float the filters on these, trying to avoid getting liquid on the colony side of the filter.

It is possible to take multiple copies (or 'lifts') from a single plate. This enables two lifts from the same plate to be hybridized to the same probe. Background spots will not normally appear in the same place on the two filters, whereas real signals should always appear in both filters. This is a quick and easy way to eliminate many false positives. Taking multiple lifts also allows the same set of clones to be probed with different probes. Bacterial colonies are transferred to nitrocellulose rather efficiently and so the plates will need to be re-grown between taking each lift. At least six and sometimes 12 lifts can be taken from a bacteriophage plate. One limitation is the peeling of the top agar layer as the plates dry. On the other hand, if the plates are not sufficiently dry before the top agarose is poured, then the top agarose may peel off with the filter, especially if multiple lifts are taken from the same plate.

2.2 Stringency of hybridization and washing

Nucleic acid strands that are not perfectly complementary can still form hybrids, but the melting temperature (T_m) decreases as the degree of mismatch increases. Hybridization can therefore be used to detect sequences closely related to those of the probe in addition to identical sequences. This approach is widely used to isolate related genes. The conditions used for hybridization and washing should be carefully chosen, taking into account the nature of the probe and the target. The conditions should be stringent enough to eliminate unwanted hybridization and yet to allow hybrids formed with the desired target to be stable. Stringency can be increased by either increasing the temperature or decreasing the ionic strength while retaining otherwise similar conditions. The melting temperature or T_m (the temperature at which

50% of a nucleic acid duplex is in single-stranded form) decreases by approximately 1 °C for each 1% mismatch. Where stringency is critical, it is always best to determine the T_m empirically, for example by washing in conditions of increasing stringency, with autoradiography between each step. The conditions for hybridization and washing should be 10–20 °C below the T_m.

Protocol 4. Hybridization to filters in aqueous solution

Equipment and reagents

- Boiling water-bath
- Ice or dry ice/ethanol bath
- Saran-wrap or cling-film
- Hybridization oven (see Section 2.2.3)
- Hybridization buffer: 7% SDS, 0.5 M sodium phosphate pH 7.2,[a] 1 mM EDTA, which is prepared by mixing: 70 g Na_2HPO_4,[b] 140 g SDS,[c] 4 ml 0.5 M EDTA

- pH 8.0, and 4 ml 85% (w/v) H_3PO_4 in 2 litres H_2O
- 20 × SSC (dissolve 174.3 g NaCl and 88.2 g sodium citrate in 800 ml of H_2O, adjust the pH to 7.0 with a few drops of 10 M NaOH, make up to 1 litre with H_2O
- 10% (w/v) SDS[c]

Method

1. Soak the filter in hybridization buffer[d]. Incubate at 65 °C for 15 min to 1 h.

2. Heat the probe[e] in a boiling water-bath for 5 min.

3. Chill the probe quickly on ice or dry ice/ethanol.

4. Add the denatured probe to the hybridization buffer. Replace the pre-hybridization buffer with the appropriate volume of hybridization buffer. This depends on the apparatus used and the number of filters. Typically 10–50 ml is used. The hybridization buffer is added to the filters either in a hybridization bottle or a plastic bag which is then heat-sealed.

5. Incubate overnight with shaking at a suitable temperature (usually 65 °C—see Section 2.2).

6. Rinse the filter twice in 0.2 × SSC/0.1% SDS at room temperature.

7. Wash the filter three times for 20 min in 0.2 × SSC/0.1% SDS[f] at 65 °C. Agitate the washing vessel throughout this process.

8. Drain the filters without allowing them to dry completely. Wrap them in plastic film, e.g. Saran-wrap or cling-film, and expose to X-ray film.[g]

[a] Molarity of Na^+.
[b] Or 134 g $Na_2HPO_4.7H_2O$.
[c] Cheap grades of SDS are adequate.
[d] Pre-hybridization is usually unnecessary except for colony lifts (*Protocol 2*) and RNA filters.
[e] See Section 2.4 for discussion about probes and Chapter 4 for details of probe synthesis.
[f] For RNA probes or target or both, use 0.2 × SSC/0.5% SDS.
[g] See Section 2.5.1.

2.2.1 Hybridization buffer

The hybridization buffer in *Protocol 4* is based on that of Church and Gilbert (9). Many other hybridization buffers have been used, generally based on SSC and SDS with the addition of a range of other reagents. The buffer of *Protocol 4* is 0.99 M Na$^+$, or equivalent to 5 × SSC for the purposes of calculating stringency. Although this buffer works well as described, for particular applications it may be worth modifying it as discussed below.

(a) Hybridization in 50% formamide at 42 °C is gentler on nitrocellulose membranes than in aqueous buffers at 65–68 °C. If required, the hybridization buffer of *Protocol 4* can be made up with 50% formamide and used at 42 °C with no change in the final result. Hybridization in 50% formamide is probably slower than in aqueous solutions by a factor of about two, so do not reduce the hybridization time below 16–24 hours when using formamide buffers.

(b) Many hybridization buffers include blocking agents such as Denhardt's reagent (100 × solution is: 10 g Ficoll 400, 10 g polyvinylpyrrolidone, 10 g BSA, H$_2$O to 500 ml—normally used at 5 × concentration in some hybridization buffers), a 1% solution of dried milk, heparin (at a final concentration of 50 μg/ml), tRNA (at 50 μg/ml final concentration), or denatured sonicated salmon sperm DNA (at a final concentration of 100 μg/ml). These are used to block the non-specific binding of the probe to the surface of the filter. We have never found this to be a problem when using *Protocol 4*. One specific form of blocking agent that is occasionally useful is the vector DNA from the clone used to make the probe. 'Purified' restriction fragments of DNA used for labelling are often contaminated with vector DNA. Thus the probe may detect vector DNA on the filter giving background signals. However, if necessary this background can be much reduced by adding denatured vector DNA to the probe before hybridization. This can be useful when hybridization is being used on a replica of a dense array of colonies or plaques. 10% Dextran sulfate and 10% polyethylene glycol are sometimes added to increase the rate of hybridization by reducing the effective volume of the hybridization. We find that these enhancing agents generally increase background problems and are not recommended, particularly with *Protocol 4*.

2.2.2 Washing conditions

The post-hybridization wash conditions used in *Protocol 4* are quite stringent, and will normally only allow near-perfect hybrids to remain. To reduce the stringency of the washes, either the ionic strength should be increased, the temperature decreased, or both. The exact conditions depend on the stringency required, which is not normally known beforehand. If necessary try reducing the hybridization temperature by up to 20 °C and then washing using

$2 \times$ SSC/0.1% SDS initially at room temperature. Check the hybridization signal by autoradiography and then if necessary wash again in $2 \times$ SSC/0.1% SDS increasing temperature by 10 °C intervals and exposing between each wash. The point at which signal disappears shows its similarity to the probe. PCR is now more often used to try to clone similar sequences (see Section 3), but low stringency hybridization to Southern blots can give useful information about sequence families.

2.3 Apparatus for the incubations

The ideal conditions for hybridization include:

- accurately controlled temperature
- minimum hybridization volume
- safe containment of radioactive solutions.

These are provided by a number of commercially available hybridization ovens. Together with their bottles for probe containment, these provide a safer and more convenient alternative to the traditional method of heat-sealing the hybridizing filters in a double layer of plastic and incubating them in a water-bath. The use of rotating glass bottles allows a large number of filters to be constantly washed with the probe in a very small (5–20 ml) volume of hybridization buffer. The increased probe concentration reduces the required hybridization time, while the bottle design removes the problem of air bubbles in the plastic bag method, and acts as shield to β emission. Post-hybridization washes are performed in the same bottles. We find the relatively large diameter bottles of the Techne hybridizer 1D system to be convenient and reliable, but they are just a little too short to accommodate our largest filters. We have recently found the Hybritube system (Gibco-BRL 10117–018) to be an effective and very inexpensive alternative.

2.4 The hybridization probe

2.4.1 RNA versus DNA

In aqueous solution, RNA–RNA hybrids are more stable than RNA–DNA hybrids, which are in turn more stable than DNA–DNA ones. This results in a difference in T_m of approximately 10 °C between RNA–RNA hybrids and DNA–DNA ones. Consequently, more stringent conditions should be used with RNA probes.

2.4.2 Oligonucleotide probes

Compared with 'normal' probes of several hundred bases in length, short oligonucleotides behave rather differently as probes. *Protocol 4* should not be followed when using such probes. These probes are commonly used to screen libraries for specific conserved motifs. Both unique and degenerate oligonucleotides can be used. Degenerate oligonucleotides are a mixture of different oligonucleotides of the same length but slightly different sequence. These are particularly useful for isolating DNAs encoding a short, known amino acid sequence derived from consideration of homologies between related proteins or by peptide sequencing of a purified protein.

The T_m of an oligonucleotide probe longer than 13 bases can be calculated using the formula originally derived by Bolton and McCarthy (10):

$$T_m = 81.5 - 16.6(\log_{10}[Na^+]) + 0.41 \, (\%G + C) - (600 \, / \, length)$$

Alternatively, for oligos shorter than 18 bases, the T_m can be estimated by the equation (11):

$$T_m = 2(A + T) + 4(G + C)$$

Hybridization is normally performed at approximately 10 °C below the T_m. This is a compromise between stability of the hybrids, the time taken for the hybrids to form, and the stringency. Whereas hybrids several hundred bases in length are essentially stable under normal washing conditions, short oligonucleotide probes can melt off when extensively washed 10 °C or so below their T_m. Hybridization is therefore performed with high molar concentrations of probe and stringent conditions, with short washes. Factors affecting the T_m are of critical importance. The following factors should be taken into consideration when designing and using oligonucleotide probes for hybridization:

- length
- mismatches
- %(G + C).

The probe should be as long as is feasible without too greatly increasing the degeneracy of the probe. This is usually limited by the sequence homology or peptide sequence on which the oligo sequence is based.

The effect of mismatches on the hybridization is hard to predict. While it is approximately true that the T_m decreases by 1–1.5 °C for each 1% mismatch the effect of a particular mismatch depends on its position in the oligo. A mismatch at one end of an oligonucleotide reduces its effective length by one base, but a hybrid with a mismatch in the middle is much less stable, not hybridizing at all under normal conditions.

The problem of mismatches can be overcome by using degenerate oligonucleotides. These can represent all the possible DNA sequences that can encode a particular amino acid sequence, so that one of the oligonucleotides in the mixture will be a perfect match. However, for a highly degenerate oligonucleotide, only a tiny proportion of the total probe will be of the correct sequence. This affects both the T_m calculations and the hybridization time. The calculation of the T_m becomes difficult with mixtures of bases. In general, for a mixture of sequences it is best to try using a hybridization temperature at least 2°C below the T_m of the most A/T-rich oligo (but see below and *Protocol 4*). Of course this may be so far below the T_m of the most G/C-rich oligonucleotide that even with a mismatch it can form hybrids that are more stable than a perfect match of the most A/T-rich one. The hybridization time is affected because of the relatively low concentration of 'perfect match' oligo in the pool. The use of large numbers of distinct oligos in a degenerate pool can be avoided in two ways:

(a) The number of distinct oligonucleotides in a degenerate mixture can be significantly reduced by using a neutral base in a position of three- or four-fold redundancy. Inosine pairs with all four conventional bases with approximately equal strength. Nevertheless, T_m calculations for inosine-containing oligos are still difficult. However, an estimate can be made by assuming that the inosine bases make neither a positive nor a negative contribution to the stability of the hybrids. Thus a T_m may be calculated for a hypothetical (shorter) oligonucleotide of the same sequence excluding the inosines.

(b) Some organisms show a highly biased codon usage or nucleotide distribution. This can be used to guess the likely sequence without resorting to highly degenerate oligos. Codon usage tables are available for a number of organisms. Note that biased codon usage varies considerably between different organisms and that CpG dyads are underrepresented in mammalian DNA. The effect of the base composition on the T_m can be minimized by performing the most stringent step in the procedure in the presence of high concentrations of quaternary alkyl ammonium salts. Such salts, e.g. tetramethylammonium chloride, bind specifically to A/T base pairs, stabilizing these hybrids. At a concentration of three molar, A/T pairs are as stable as G/C ones. Under such conditions, the dissociation temperature is dependent primarily on the length of the oligo, rather than on its base composition.

Table 3. Dissociation temperature in 3 M tetramethylammonium chloride

Length of oligonucleotide (bp)	Dissociation temperature (°C)
11	44–45
13	47
15	49–50
16	53–54
18	57–58
27	70–71
31	74–76

Washing is usually best performed a few degrees below this temperature, unless it is necessary to discriminate between the hybridization of two oligos of similar length, or between a perfect match and a mismatch at one end of the oligo.

Protocol 5. Oligonucleotide hybridization using tetramethylammonium chloride

Equipment and reagents

- Saran-Wrap or cling-film
- X-ray film and cassette
- 20 × SSC (see *Protocol 4*)
- 50 × Denhardt's solution: 5 g Ficoll 400 (Pharmacia), 5 g polyvinylpyrrolidone, 5 g bovine serum albumin in a final volume of 500 ml H$_2$O
- Pre-hybridization buffer: 6 × SSC, 50 mM

sodium phosphate pH 6.8, 5 × Denhardt's solution, 100 μg/ml boiled sonicated salmon sperm DNA
- Hybridization buffer: pre-hybridization buffer with 100 mg/ml dextran sulfate
- Wash buffer: 3 M tetramethylammonium chloride, 50 mM Tris–HCl pH 8.0, 2 mM EDTA, 0.1% SDS

Method

1. Pre-hybridize filters in pre-hybridization buffer for at least 2 h at 37 °C. Alternatively, use the hybridization buffer from *Protocol 4* for both pre-hybridization and hybridization.

2. Pre-warm hybridization buffer to 37 °C.

3. Add the oligonucleotide probe (*see Chapter 4* for method of labelling) to pre-warmed hybridization buffer to a concentration of 180 pM. Replace pre-hybridization buffer with an appropriate volume of hybridization buffer. This depends on the number of filters and the apparatus used. Typically 20 ml is used for up to ten 60 mm filters.

4. Incubate at 37 °C overnight, or longer for very degenerate probes.

5. Rinse the filters three times with 6 × SSC at room temperature.

6. Wash the filters twice for 30 min in 6 × SSC at room temperature.

7. Rinse the filters in wash buffer at room temperature.

8. Wash the filters twice for 20 min in wash buffer at the required temperature (see *Table 3* and text).

9. Rinse filters in 6 × SSC.

10. Wrap the damp filters in plastic film, such as Saran-Wrap or cling-film, and expose to an X-ray film (see Section 2.5).

2.5 Detecting the signal

How the signal is detected depends on what label has been incorporated into the probe. Probes labelled with ^{32}P are detected by autoradiography; non-radioactive probes are detected with methods and reagents specific to the label. These usually involve binding an enzyme such as peroxidase or alkaline phosphatase to the label and using this enzyme in a colour reaction which deposits a coloured precipitate at the site of the probe (12). An extensive discussion of the use of non-radioactive labels in the context of filter hybridization is beyond the scope of this chapter. However, a protocol using non-radioactive detection to study the distribution of transcripts in tissue *in situ* is given in Section 4.

Autoradiography is commonly used to detect hybridization signals on filters. Filters should not normally be dried completely before autoradiography. Dry filters may irreversibly bind the probe, which is undesirable if the filter is to be reused, or if another, more stringent wash may be required. Moreover, nitrocellulose filters become brittle when dry. The plastic film used to wrap damp filters is thin enough not to block the β-emission from ^{32}P, while keeping them from direct contact with the film. X-ray film cassettes are used to hold the filters pressed flat against the film and to seal them safely from the light. Intensifying screens can also be used to give an approximately fivefold increase in sensitivity when the film is exposed at $-70\,°$C. The film is sandwiched between the filters and the intensifying screen. β-Particles which are absorbed by the film give a signal, those which pass through are absorbed by the intensifying screen which emits a flash of light, which is then detected by the film. Several companies supply X-ray film, autoradiography cassettes, and intensifying screens. The exact procedure for developing the film depends on the source of the film—follow the manufacturers' instructions.

2.6 Reusing filters

If the signal strength on a filter is weak, or if the probe has decayed since the filter was last hybridized the filter can be reused immediately. Note that some of the DNA will be double-stranded and not available for hybridization if it is still annealed to the old probe. In other cases radioactive probe may still be present on the filter. In either of these events it will be necessary to strip the probe from the filter. This is done by one of the following methods:

- boil for 5–20 min in 0.1% SDS
- incubate in 50 mM Tris–HCl pH 8.0, 60% formamide and 1% SDS for 1 h at 75 °C.

These methods are essentially equivalent to washing at such a high stringency that no hybrids can remain. The second method is recommended for northern blots as RNA is more readily hydrolysed by high temperature or alkaline conditions. Unfortunately, the probe is rarely removed completely. The extent to which it has been removed can be monitored by exposing the filters before and after stripping and comparing the results.

3. PCR as an alternative to hybridization

The polymerase chain reaction (PCR) is a rapid procedure for the *in vitro* enzymatic amplification of a specific segment of DNA. The number of applications for which the PCR can be used is continuously growing. We have included a basic protocol which may be used in some instances as an alternative to Southern blotting. A detailed discussion of all the parameters that may affect the outcome of a PCR reaction is beyond the scope of this chapter and the reader should refer to any one of a number of excellent texts devoted to PCR, including a volume in this series (13).

Degenerate oligonucleotide primers may be used as an aid to cloning genes where only protein sequence is available or to isolate homologues of a gene in other species. PCR primers may also be designed against conserved sequences within members of a gene family to identify new genes belonging to it. When designing oligonucleotide primers the following points should be borne in mind:

(a) Try to minimize the likelihood of mismatches at the 3′ end of the primer. If the primer is designed against a peptide sequence or alignments of peptide sequences the 3′ end should ideally be based on amino acids which only have a single codon (methionine and tryptophan). Amino acids such as arginine or serine, which show degeneracy at more than one position within the codon, should be avoided if at all possible when designing a DNA primer against a protein sequence (or consensus sequence).

(b) Ideally primers should be 20–30 bases in length. It is inadvisable to attempt to use primers which are shorter than 17 nucleotides in length.

(c) Many organisms show preferential usage of certain codons. Consideration should be given to this when designing primers. Codon usage tables are available for most organisms.

(d) Check the primer sequence for any potential secondary structure. Polypurine or polypyrimidine stretches should be avoided. Additionally, the primer sequences should be checked for any complementarity to each other (especially at their 3′ ends) to avoid the possibility of 'primer-dimers' forming during the PCR reaction.

(e) Inosine may be incorporated into the primer sequence at positions of high degeneracy. Inosine may assist in primer stability during the early rounds of amplification. No more than four inosines should be incorporated in each stretch of 16 bases and the inosines should be fairly evenly distributed throughout the primer sequence. A mixture of bases may be incorporated into the oligonucleotide sequence at positions where the genetic code shows degeneracy.

(f) Cleavage sites for restriction endonucleases may be incorporated at the 5′ end of a primer. The inclusion of these sites helps to facilitate the cloning of PCR products and assists in the stability of the primer binding to the target DNA in later rounds of amplification.

It should be noted that well-designed primers against conserved stretches of amino acids does not guarantee success in the PCR reaction. Before embarking on a PCR reaction in which degenerate oligonucleotides are used it is worthwhile ensuring that the reagents to be used work. If possible a control PCR reaction should be set-up using primers which hybridize to a known DNA sequence within the template DNA being used. Because the concentration of Mg^{2+} ions in the buffer can greatly influence the outcome of the PCR this should be optimized for each set of primers to be used. The optimum Mg^{2+} concentration should fall within the range 0–10 mM. A seemingly small difference in the concentration of Mg^{2+} can make a huge difference in the outcome of a PCR reaction.

A basic PCR protocol is given in (*Protocol 6*). The optimum conditions for any primer combination must be determined empirically.

Protocol 6. Polymerase chain reaction

Equipment and reagents

- Automated thermal cycler (or water-baths if not available)
- 100 mM $MgCl_2$
- 10 × amplification buffer: 100 mM Tris–HCl, 500 mM KCl, pH 8.3 at 20 °C
- 5 mM dNTP mix: 20 mM dATP, dCTP, dGTP, and dTTP mixed in a 1:1:1:1 ratio

- 25 μM PCR primers 1 and 2 (store in sterile H_2O at −20 °C)
- 50 μg/ml template DNA
- 2.5 U/μl *Taq* DNA polymerase (Boehringer-Mannheim)
- Sterile H_2O

Method

1. Prepare an amplification mixture by adding the following reagents to a sterile 0.5 ml microcentrifuge tube:
   - 10 μl   10 × PCR buffer
   - 4 μl   5 mM dNTP mix
   - 0–10 μl   100 mM $MgCl_2$ [a]
   - 2 μl   primer 1
   - 2 μl   primer 2
   - 2 μl   template DNA [b]
   - 100 μl   final volume with sterile H_2O

   When preparing several reactions a cocktail may be made up of the common components before dispensing equal aliquots into each tube.

2. Add 1 μl of 2.5 U/μl *Taq* DNA polymerase to each tube.

3. Add a drop of mineral oil to the top of the mixture in each tube to reduce evaporation. The mineral oil should completely cover the surface of the reaction components.

4. Heat samples to 95 °C for 90 sec.

5. Incubate samples for 2 min at 50 °C.[c]

6. Incubate samples for 2 min at 72 °C.[d]

7. Repeat steps 4–6 for another 25 to 35 cycles.

8. Analyse reaction products (use 10 μl from each tube) on a non-denaturing polyacrylamide gel or agarose gel appropriate to the size of the reaction products expected.

[a] A range of $MgCl_2$ concentrations should be tried. The concentration of $MgCl_2$ in the buffer is a variable that can greatly influence the outcome of a PCR reaction.

[b] The amount of DNA required will vary according to the complexity of the genome being analysed. 100 ng of mammalian genomic DNA is usually sufficient to detect a single copy gene after 30 rounds of amplification.

[c] When using 'guessmer' oligonucleotides as primers it is often difficult to estimate a suitable annealing temperature to use. Using a low annealing temperature for the first two cycles (e.g. 37 °C) and then increasing the temperature of the annealing reaction near to the estimated T_m of the oligonucleotide–DNA hybrid in later cycles often helps. Consideration should be given to the degeneracy of the oligonucleotides when calculating the T_m.

[d] The extension time may be varied according to the expected length of the PCR product. 30 sec is normally sufficient for products up to 400 bp in length. Allow roughly 1 min for each 1000 bp of expected product size. 72 °C is close to the optimum working temperature for the *Taq* DNA polymerase.

The polymerase chain reaction is capable of amplifying as little as a single molecule of DNA. Therefore precautions should be taken to guard against contamination of the reaction mixture with trace amounts of DNA that can act as a template. We have generally found it unnecessary to set aside an area of the laboratory with equipment used solely for setting-up PCR reactions. To avoid the possibility of contamination with trace amounts of DNA it is best to store all solutions etc. frozen in small aliquots at −20 °C.

It is vitally important to include several controls in each set of PCR reactions:

(a) A positive control. Primers are designed against a known gene contained within the genomic (or other DNA that is used as a target sequence).

(b) A negative control. The reaction mixture should contain all components except for the template DNA.

(c) A single primer control. Every primer should be tested in the absence of its partner. A single primer (particularly when it is a 'guessmer oligonucleotide') can often give rise to a strong amplified signal. These controls can help in interpreting results when multiple bands are observed after a PCR reaction.

Any products observed on a gel after the first round of PCR amplifications should be isolated and re-amplified under high stringency conditions. The products of the first round of amplifications should include sequences which are exactly complementary to the oligonucleotides used at their 5′ and 3′

ends. Therefore the annealing temperature used should be very near to the calculated T_m for primer–target DNA hybrids (see Section 2.4.2 for details of how to calculate the T_m). The re-amplified products may then be used for sequencing or for use as a probe in screening a cDNA or other DNA library. Any sequence obtained may be used to design new PCR primers for use in anchored PCR to sequence a full-length copy of the genes transcript (see Section 5.4).

4. *In situ* hybridization

In situ hybridization allows specific nucleic acid sequences to be detected in intact cells, chromosomes, or tissue sections. This powerful approach has many applications, including the mapping of genes to chromosome regions, the analysis of gene activity through development at the cellular or sub-cellular level, and the detection of viral nucleic acids in tissue biopsies. However, specialized protocols are required for each application and specimen type, so only general comments will be presented here. Full protocols for two typical applications (hybridization to *Drosophila* polytene chromosomes and hybridization to RNA of *Drosophila* embryos) are provided in ref. 14. A range of protocols using digoxigenin-based probes are available from Boehringer-Mannheim (15).

4.1 Choice of probe and detection of the labelled nucleic acid

Accurate localization of nucleic acids *in situ* requires that the final signal is detected at the site of the target DNA. This means that non-radioactive probes are far preferable to radioactive ones, as even a low energy emitter such as tritium can only be detected as silver grains some distance from the actual target DNA. Non-radioactive methods can be divided into direct and indirect methods. In a direct method, the labelled nucleic acid can be directly detected microscopically, for example by incorporating a fluorochrome. Indirect methods require an additional step in which the labelled nuclei acid is detected by a reporter molecule, for example a biotin labelled probe detected with a labelled streptavidin. Indirect detection methods allow the signal to be amplified, but usually at some cost in background and time. Similarly, a conjugated enzyme (e.g. alkaline phosphatase or horse-radish peroxidase) can be used as the reporter molecule and visualized by using the enzyme to catalyse a reaction with coloured or fluorescent products. This amplification of the original signal gives a great increase in sensitivity, but again with some cost in background and time, and also accuracy, as the reaction products will diffuse somewhat. Non-radioactive probes are discussed in more detail in Chapter 4, Section 5.

Although the hybridization rate in solution is greater for longer probes, for *in situ* hybridization there is the problem of probe penetration into the fixed specimen. As a compromise, probe lengths of about 150–300 bases are used. This means that random primed probes (see Chapter 4, *Protocol 1*) can be used without further treatment. *In vitro* transcribed RNA probes (Chapter 4, *Protocol 3*) may need size reduction by partial alkaline hydrolysis, e.g. with carbonate buffer. The advantages of single-stranded probes are discussed in Chapter 4.

4.2 Fixation

The specimen must be fixed to preserve morphology. A wide-range of fixatives can be used, but a post-fixation permeabilization step is generally required as the nucleic acid target sequences may be masked by fixed proteins. In whole mount procedures, the greater permeabilization will allow target sequences to be detected deeper within the specimen, but usually at the expense of some morphological preservation. Unfortunately, no single fixation procedure is suitable for all specimens, and this remains the most variable step between different protocols.

4.3 Hybridization

As for hybridization of nucleic acids in solution, the main parameters affecting the hybridization rate are:

- temperature
- concentration of probe
- concentration of monovalent cations, e.g. Na^+
- presence of organic solvents
- pH.

As for hybridization to nucleic acids immobilized on solid supports, neutral pH (6.5–7.5) and high Na^+ concentrations (0.4–1.0 M) are used. Temperature is more of a problem. Hybridization in aqueous solution requires prolonged incubation at 65–70 °C (see *Protocol 4*), which will usually have an adverse effect on the morphology of the specimen. The answer is to add formamide to 50% or more, as discussed in Section 2.2.1. Strong detergents such as SDS may also adversely affect the specimen, non-ionic detergents such as NP-40 or Tween-20 are more appropriate. Background is more of a problem for *in situ* hybridization than for filter hybridization, so most protocols employ one or more of the background reducing agents discussed in Section 2.2.1. A pre-hybridization step is often required to reduce background hybridization. The pre-hybridization solution is the same as the hybridization mix but omitting the probe.

5. Transcript mapping techniques

Northern blotting (see Section 2.1.2) is useful for determining the size and amount of specific mRNA molecules. However, little information is gained about the gene structure. Below several techniques are described that enable the precise 5′ and 3′ ends of an mRNA molecule to be determined and the location of any splice sites within it.

5.1 Mapping RNA with nuclease S1

Nuclease S1 mapping is used to map the locations of the ends of RNA molecules and of any splice junctions within them in relation to specific sites (e.g. positions of restriction endonuclease cleavage) within the template DNA (16). Hybridization conditions can be established that minimize the formation of DNA–DNA hybrids whilst promoting the formation of DNA–RNA hybrids. In high concentrations of formamide, DNA–RNA duplexes are more stable than the equivalent DNA–DNA duplex. When double-stranded DNA is denatured and hybridized with RNA under these conditions regions of the DNA molecule homologous to those in the RNA form stable hybrids. Any non-homologous regions in the DNA molecule (e.g. introns) will remain single-stranded. DNA that forms stable duplexes with the RNA will be protected from digestion with nuclease S1 whereas any single-stranded regions will be digested. The principles of S1 mapping are shown in *Figure 2* for a hypothetical eukaryotic gene containing one intron. In an S1 mapping experiment probes may be either 5′ or 3′ end-labelled, uniformly labelled, or unlabelled (see Chapter 4). If an unlabelled probe is used the electrophoretic-ally separated samples may be detected by means of Southern blotting (*Protocol 1*). 5′ End-labelling is useful for determining the 5′ end of an RNA and

---

Figure 2. Nuclease S1 mapping of RNA. The *top* panel of the diagram shows the DNA structure of a hypothetical eukaryotic gene containing one intron. Positions of restriction endonuclease cleavage are indicated as follows: B = *Bam*HI; H = *Hin*dIII; R = *Eco*RI. In the experiment shown several probes are used. In the experiment depicted on the *left* side, a 7.0 kb genomic probe (*Bam*HI–*Bam*HI fragment) is used. Hybrids formed between the transcribed strand of genomic DNA and mRNA contain a single-stranded loop of DNA (the intron). Digestion with nuclease S1 divides the probe into 1.5 kb and 2.0 kb segments revealing the sole intron to be near the centre of the coding region. In the experiment shown on the *right* side, multiple probes are used. The 2.0 kb *Bam*HI–*Hin*dIII fragment contains the 5′ end of the transcription unit. Upon S1 digestion of hybrids formed between this probe and the RNA a 1.0 kb fragment is protected revealing the start site of transcription to be 1.0 kb upstream of the *Hin*dIII site. The 1.5 kb *Eco*RI–*Bam*HI fragment contains the 3′ end of the transcription unit. A 0.5 kb fragment is protected after S1 digestion revealing the end of the mRNA to be 500 nucleotides downstream of the *Eco*RI site. When the 3.5 kb *Hin*dIII–*Eco*RI probe is used, 1.5 kb and 0.5 kb bands are observed after gel electrophoresis. When the information obtained using all four probes is put together the location and size of the intron can be derived and the 5′ and 3′ ends of the RNA can be mapped on to the genomic DNA.

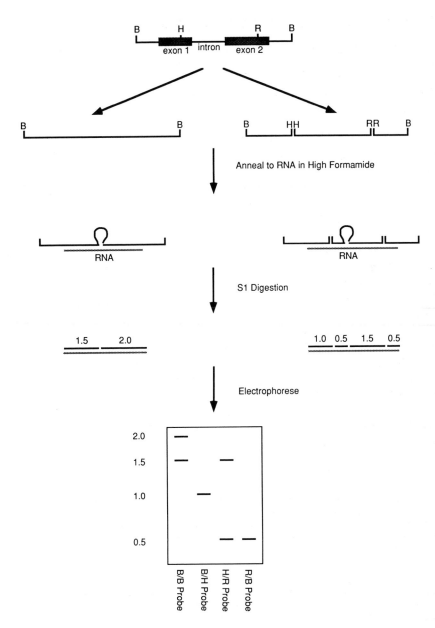

the position of 3′ splice sites. Conversely 3′ end-labelled probes may be used to map 3′ ends of RNA or 5′ splice sites. A uniformly labelled DNA or RNA probe is most commonly employed and will have a higher specific activity than 5′ or 3′ end-labelled probes.

When using double-stranded DNA probes the hybridization temperature is

165

Table 4. Hybridization temperatures for S1 mapping

G + C	Hybridization temperature
41%	49 °C
49%	50 °C
58%	55 °C

chosen to minimize the formation of DNA–DNA hybrids while allowing DNA–RNA hybrids to form. The hybridization temperature depends on the G + C content of the DNA (*Table 4*). The optimal hybridization temperature is likely to fall within this range. However, it is advisable to carry out preliminary experiments to determine the optimal hybridization conditions for the RNA being used.

The conditions for nuclease S1 digestion should be determined for each RNA–DNA hybrid. Although nuclease S1 is most active at 45 °C, incubation at 20 °C will minimize the digestion of RNA at the base of DNA loops (e.g. formed by intron sequences which have been spliced out of the mRNA). When these molecules are run on neutral gels under non-denaturing conditions, the RNA remains hybridized to the DNA and a single band will be seen. If the same sample is run on an alkaline gel under denaturing conditions the RNA bridge will by hydrolysed and two smaller pieces of single-stranded DNA will be liberated. Neutral gel analysis can often assist in the assignation of intron/exon boundaries (16).

Protocol 7. Nuclease S1 mapping of RNA

Reagents

- Hybridization buffer: 40 mM Pipes pH 6.4,[a] 1 mM EDTA pH 8.0, 0.4 M NaCl, 80% formamide[b]
- S1 buffer: 0.28 M NaCl, 0.05 M sodium acetate pH 4.6, 4.5 mM $ZnSO_4$, 100–3000 U/ml nuclease S1
- Phenol/chloroform (25:24:1 phenol/chloroform/isoamyl alcohol)
- 3 M sodium acetate pH 5.2
- Carrier RNA (e.g. tRNA)
- Formamide dyes: 0.05% bromophenol blue, 0.05% xylene cyanol in 95% formamide
- Radiolabelled markers (e.g. a *Hae*III restriction digest of φX174 DNA which has been end-labelled as described in Chapter 4)

Method

1. Mix in a microcentrifuge tube 10^4 c.p.m. (Cerenkov) of a 5′ or 3′ end-labelled DNA probe labelled to a specific activity of $1–2 \times 10^6$ c.p.m./pmol end. For uniformly labelled probes use $> 5 \times 10^4$ c.p.m. and not less than 5 fmol (the specific activity should be $> 5 \times 10^7$ c.p.m/g). Add 1–100 µg RNA (the amount of RNA used depends on the abundance of the message).

2. Add 0.1 vol. of 3 M sodium acetate pH 5.2 and 2.5 vol. ethanol. Mix and spin in a microcentrifuge at full speed for 20 min. Discard the supernatant and wash the pellet with 70% ethanol. Re-centrifuge for 5 min. Carefully remove the supernatant and allow the pellet to air dry (do not allow pellet to become desiccated).

3. Dissolve in 30 µl of hybridization buffer.

4. Denature the nucleic acids in a water-bath set to 90 °C for 10–15 min.

5. Rapidly transfer the tubes to a water-bath set to the desired hybridization temperature (see text). Do not allow the tubes to cool. This can be ensured by carrying out the incubation with the sealed tubes submerged in the bath.

6. Hybridize for at least 3 h (preferably overnight).

7. Add 270 µl of cold S1 buffer. Incubate at the appropriate temperature for 30–120 min (see text).

8. Stop the reaction and extract the nucleic acids by the addition of an equal volume of phenol/chloroform. Vortex and spin for 3 min in a microcentrifuge. Transfer the supernatant to a fresh tube.

9. Add 2 µg of carrier RNA. Precipitate with 2.5 vol. ethanol and wash the pellet (as in step **2**, except that the addition of sodium acetate is not necessary).

10. Re-dissolve the pellet in 10 µl of formamide dyes (or 50% glycerol plus 0.2% bromocresol green if running a neutral gel). Heat the samples to 95 °C for 5 min, and transfer immediately to ice.

[a] Use the disodium salt of Pipes.
[b] Use formamide deionized by passage through a mixed-bed resin (e.g. Bio-Rad AG 501–X8), filtered through Whatman No. 1 paper, re-crystallized at 0 °C, and stored in small aliquots at −70 °C.

5.2 RNase mapping

RNase mapping is analogous to S1 mapping (17), but uses RNA as labelled probe prepared as in Chapter 4, *Protocol 3*, and RNases A and T1 to digest regions of the probe that do not form hybrids with the corresponding gene or mRNA. Under the conditions used these enzymes cleave only single-stranded RNAs. Mapping of RNAs with RNA probes is extremely sensitive and less than 1 pg of mRNA can be detected. An outline procedure for RNase mapping is illustrated in *Figure 3*.

The optimal conditions for forming RNA–RNA hybrids varies from RNA to RNA and is influenced by factors such as the ability to form secondary structures and the G + C content. A hybridization temperature of 45 °C

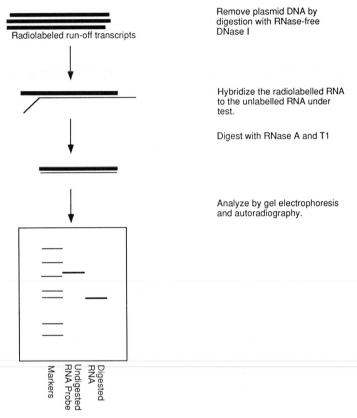

Figure 3. RNase mapping of RNA. The probe RNA is made as illustrated in Chapter 4, *Figure 2*. The radiolabelled RNA is hybridized to the unlabelled RNA under test and digested with RNases A and T1. The resulting digestion products are analysed by gel electrophoresis and autoradiography.

works well in most cases. Should this not be the case a range of hybridization temperatures between 25 °C and 65 °C may be tried.

Optimal conditions may be established in test experiments in which the labelled probe is hybridized to RNA made from the opposite strand of the DNA template. The digestion of RNA–RNA hybrids with RNase suffers from fewer artefacts than digestion of DNA–RNA hybrids with nuclease S1 and is normally the method of choice for quantitating mRNA molecules, mapping their ends, and determining the position of introns within the gene.

The optimal time and temperature of digestion by RNases may have to be established for each probe. A one hour incubation at 30 °C is usually sufficient using the conditions given in *Protocol 8* but if high background develops, higher temperatures or longer periods of digestion may help. High background signals may also result from using too much probe, or from the incomplete digestion of the DNA template in the preparation of the probe. If quantitative results are required the probe should be in excess (10- to 20-fold) over the complementary sequences being analysed.

Protocol 8. Mapping of RNA with uniformly labelled RNA probes and ribonuclease

Reagents

- RNase digestion mixture: 300 mM NaCl, 10 mM Tris–HCl pH 7.5, 5 mM EDTA pH 7.5 2 µg/ml RNase T1,[a] 40 µg/ml RNase A[b]
- 10% SDS (w/v)
- Proteinase K[c]

- Phenol/chloroform (25:24:1 phenol/chloroform/isoamyl alcohol)
- Carrier RNA (e.g. tRNA)
- Formamide dyes (*Protocol 7*)
- Radiolabelled markers (*Protocol 7*)

Method

1. Mix in a microcentrifuge tube 1–100 µg of the RNA to be analysed (the amount depending on the abundance of the transcript being analysed). Add 5×10^5 c.p.m. of high specific activity RNA probe (prepared as in Chapter 4, *Protocol 3*. The template DNA must be completely removed by treatment with RNase-free pancreatic DNase I (step 7).

2. Follow *Protocol 7*, steps 2–4.

3. Quickly transfer the hybridization mixture to a water-bath set at 45 °C (see text).

4. Incubate for 8–16 h.

5. Transfer to room temperature and add 300 µl of RNase digestion mixture.

6. Incubate for 1 h at 30 °C.

7. Add 20 µl of 10% SDS and 50 µg of proteinase K.[c] Incubate the mixture for 15 min at 37 °C.

8. Add 350 µl of phenol/chloroform and mix on a vortex machine. Centrifuge for 3 min. Transfer the upper phase to a clean tube.

9. Add 5 µg of tRNA (carrier) and 2.5 vol. ethanol. Mix well. Concentrate the precipitate and wash the pellet by re-centrifugation as in step 2.

10. Carefully remove all of the ethanol and air dry the pellet until no traces of ethanol can be seen in the tube.

11. Dissolve the pellet in 10 µl formamide dyes (see *Protocol 7*). Heat to 95 °C for 10 min and then transfer to an ice bucket. Analyse by electrophoresis on a polyacrylamide/7 M urea gel together with radiolabelled markers.

[a] Commercial preparations of RNase T1 are usually pure and do not need to be boiled. Make up a 100 g/ml stock solution in H_2O. Store at −20 °C.

[b] Prepare a 10 mg/ml solution in H_2O. Boil for 10 min. Allow to cool to room temperature and then store at −20 °C.

[c] Make up a 10 mg/ml stock solution of proteinase K immediately before use. Stored solutions of proteinase K do not work well.

5.3 Primer elongation

This technique can be used to map the precise 5' end of an RNA molecule. Typically a 5' end-labelled oligonucleotide (see Chapter 4), or short restriction fragment, is hybridized to a poly(A)$^+$ RNA sample. The binding site for the primer should be as close to the 5' end of the RNA as possible (ideally within 100 nucleotides). In the assay the primer is extended using reverse transcriptase and nucleoside triphosphates (see *Figure 4*). When the 5' end of the RNA molecule is reached the reaction terminates. The primer extension products are detected by polyacrylamide gel electrophoresis. Both the labelled primer and the products of the extension reaction can be seen after autoradiography. A series of weaker bands, migrating faster than the fully extended product, are usually seen. These arise as reverse transcriptase often pauses at secondary structures in the RNA being assayed. The same oligonucleotide primer can be used in DNA sequencing reactions performed on cloned genomic DNA of the gene of interest. The products of the sequencing reactions when run alongside the elongated primer can be used to position the exact mRNA start site in relation to the DNA sequence. Any heterogeneity in start site selection can also be mapped. This technique is particularly useful for detecting any heterogeneity in start site selection when studying the effects of mutations made in a gene promoter. When performing an *in vitro*

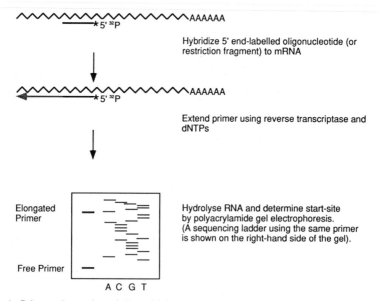

Figure 4. Primer elongation. A 5' end-labelled oligonucleotide primer is hybridized to the RNA under test. The 5' end of the RNA is mapped by using the primer to synthesize a complementary DNA strand. DNA synthesis is halted when the end of the RNA molecule is reached.

transcription reaction, primer elongation is often a good way of visualizing the transcribed RNA products.

Protocol 9. Mapping the 5′ end of RNA molecules by primer elongation

Reagents

- Dimethylsulfoxide (DMSO)
- RNA carrier (e.g. total yeast RNA)
- 2 × annealing buffer: 100 mM Tris–HCl pH 8.3 (measured at 42 °C), 10 mM MgCl$_2$, 100 mM KCl, 2 mM dithiothreitol (make up just prior to use)
- 5′ End-labelled oligonucleotide (approx. 50 000 c.p.m./μl)
- Propan-2-ol
- 2.5 mM dNTPs: 10 mM dATP, 10 mM dCTP, 10 mM dGTP, 10 mM dTTP, mixed in a 1:1:1:1 ratio
- Reverse transcriptase (e.g. AMV reverse transcriptase from Boehringer-Mannheim)
- 10 M stock solution of NaOH
- 1 M Tris–HCl pH 8.0
- HCl
- Phenol/chloroform (25:24:1 phenol/chloroformso/amyl alcohol)
- Formamide dyes (*Protocol 7*)
- Radiolabelled markers (*Protocol 7*) or a sequencing reaction performed on cloned genomic DNA using the primer to be used in this experiment (see Chapter 7)

Method

1. RNA extracted from tissue or cells in culture is resuspended in 90% DMSO (v/v) to give a final RNA concentration of at least 100 ng/μl. Typically 10 μg of RNA is dissolved in 100 μl of 90% DMSO. If low amounts of RNA are to be used add 10 μg of RNA carrier.

2. Incubate for 20 min at 45 °C. Add 0.1 vol. of 3 M sodium acetate pH 5.2 and 2.5 vol. ethanol. Mix and spin for 15 min in a microcentrifuge.

3. Resuspend the RNA pellet in 30 μl of reverse transcriptase buffer (15 μl of 2 × annealing buffer, 2 μl of the labelled oligonucleotide, 13 μl of H$_2$O).

4. Incubate at 55 °C for 30 min. Allow the samples to cool slowly to 42 °C (usually by switching off the water-bath and waiting until the temperature drops to 42 °C).

5. Add 0.2 vol. of 3 M sodium acetate pH 5.2 and 0.6 vol. of propan-2-ol. Mix and centrifuge for 10 min in a microcentrifuge. Carefully remove the supernatant.

6. Take up the pellet in 40% propan-2-ol and 0.3 M sodium acetate pH 5.2. Mix and centrifuge for 10 min. Remove the supernatant and air dry the pellet.

7. Resuspend the pellet in:
   - 15 μl of 2 × annealing buffer
   - 12 μl of 2.5 mM dNTP mixture
   - 20 U (usually 1 μl) of reverse transcriptase
   - H$_2$O to a total volume of 30 μl.

171

Protocol 9. *Continued*

   8. Incubate for 2 h at 42 °C.

   9. Hydrolyse the RNA by adding NaOH to a final concentration of 0.3 M. Incubate for 10 min at 56 °C.

10. Neutralize by adding Tris–HCl pH 8.0 to 0.62 M and HCl to 0.22 M (final concentrations). Add carrier RNA to a final concentration of 100 ng/µl.

11. Add an equal volume of phenol/chloroform. Mix by vortexing and centrifuge for 3 min.

12. Transfer the upper phase to a new tube. Add 0.1 vol. of 3 M sodium acetate pH 5.2 and 2.5 vol. ethanol. Mix well and spin for 15 min in a microcentrifuge. Carefully remove the supernatant.

13. Rinse the pellet twice with 70% ethanol, dry briefly, and resuspend in 4 µl of formamide dyes. Heat the samples to 90 °C for 5 min and analyse by electrophoresis on a sequencing gel together with the products of the sequencing reaction performed using the same primer.

5.4 Anchored polymerase chain reaction

A wide variety of techniques have been described in the literature detailing how PCR can be used to amplify segments outside boundaries of known sequence (18). Many of these techniques involve the creation of new primer binding sites by the enzymatic addition of a 'tail' or by the ligation of an oligonucleotide cassette to the end of a DNA fragment. Here we describe a single technique that has proved useful in generating copies of a specific cDNA between a point of known sequence and either end. This technique called the 'rapid amplification of cDNA ends' (RACE) is a form of 'anchored' PCR and has been described in a series of papers by M. A. Frohman and his collaborators (19, 20).

RACE is a powerful technique utilizing PCR, and can now complement the transcript mapping techniques described above. Frequently cDNA libraries contain a large proportion of prematurely terminated cDNA molecules and rare messages may be underrepresented in any given cDNA library. Providing a small amount of sequence information is available it is now possible to amplify a target cDNA where sequences either side of the sequenced region are unknown. RACE can also be used to follow up the PCR reactions described in Section 3 providing some sequence information is obtained on the amplified fragment.

5.4.1 Amplification of regions downstream of a known sequence

Three PCR primers are required in order to amplify sequences 3′ to the known region of DNA: a 20mer oligo(dT) primer and two specific primers (see *Figure 5*). Initially mRNA is reverse transcribed using the oligo(dT)

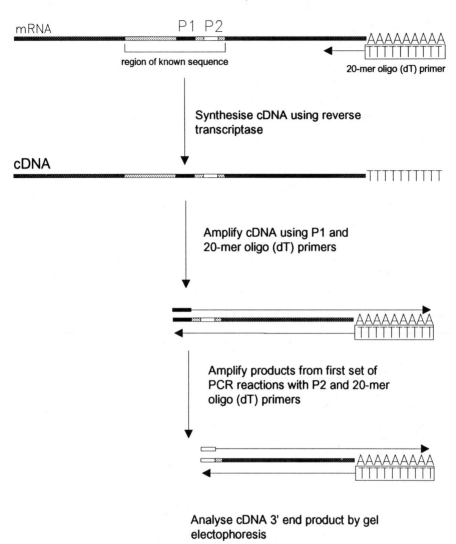

Figure 5. Amplification of the 3′ ends of cDNA by anchored PCR. Initially a pool of RNA is reverse transcribed using a 20mer oligo(dT) primer. The cDNA library generated is then amplified using a gene-specific primer (P1) contained within a region of known sequence of the gene of interest and the same 20mer oligo(dT) primer used for the reverse transcription reaction. The products from the first set of PCR reactions are then re-amplified using a second gene-specific primer, P2, located downstream of the 3′ end of P1 and the 20mer oligo(dT) primer. The use of a different primer for re-amplification helps to ensure that only specific products are amplified. After the second set of PCR amplifications a single band should be visible on an ethidium bromide stained agarose gel.

primer. PCR amplification is then carried out using a specific primer (P1 in *Figure 5*) and the oligo(dT) primer. The efficiency of the reverse transcription reaction is relatively unimportant as any cDNAs that do not extend past the binding site of the PCR primer will not participate in the subsequent amplifications. A second round of amplification cycles are then carried out using a primer (P2 in *Figure 5*) that is downstream of the first primer (P1) and the oligo(dT) primer. The protocol described here is a slight variation on that described by M. Frohman (20).

Protocol 10. Amplification of the 3′ end of a cDNA molecule

Equipment and reagents

- Automated thermal cycler
- Source of RNA
- Moloney murine leukaemia virus (M-MLV) reverse transcriptase (200 U/μl)
- 2.5 U/μl *Taq* DNA polymerase (Boehringer-Mannheim or Perkin-Elmer Cetus)
- 25 pmol/μl each of sequence specific primers (P1 and P2)
- 20 pmol/μl oligo(dT)$_{20}$ primer

- 10 × amplification buffer: 670 mM Tris–HCl pH 9.0, 67 mM MgCl$_2$, 1.7 mg/ml bovine serum albumin, 166 mM (NH$_4$)$_2$SO$_4$
- 5 mM dNTP mix: 20 mM dATP, dCTP, dGTP, dTTP, mixed in a 1:1:1:1 ratio
- RNasin (Promega)
- Mineral oil
- TE: 20 mM Tris-HCl pH 8.0, 1 mM EDTA

Method

1. Assemble the reverse transcription components on ice:
   - 4 μl 5 × reverse transcription buffer (as supplied by manufacturer of enzyme)
   - 4 μl dNTP mix
   - 0.5 μl (10 U) RNasin
   - 0.5 μl (10 pmol) oligo(dT)$_{20}$ primer.

2. Heat 1 μg of poly(A)$^+$ RNA or 5 μg of total RNA at 90 °C for 3 min. The RNA should be in 10 μl of H$_2$O.

3. Centrifuge the RNA briefly and then chill on ice prior to adding it to the reaction mix from step **1**.

4. Add 1 μl (200 U) of reverse transcriptase and incubate the reaction mixture for 2 h at 37 °C. This results in the synthesis of cDNAs.

5. Dilute the reaction mixture to 1 ml with TE and store at 4 °C.

6. Prepare the following PCR reaction on ice:
   - 1 μl cDNA template (from step **5**)
   - 10 μl 10 × amplification buffer
   - 1 μl (25 pmol) of primer 1 (see *Figure 5*)
   - 1.25 μl (25 pmol) of oligo(dT)$_{20}$ primer

- 4 μl dNTP mix
- 10 μl DMSO
- 72.75 μl H_2O.

7. Heat sample in a thermal cycler or water-bath at 95°C for 5 min to denature the cDNA products.

8. Cool to 75°C. Add 2.5 U of *Taq* polymerase and overlay with one drop of mineral oil.

9. Incubate at 42°C for 2 min.

10. Extend the cDNAs at 72°C for 40 min.

11. Carry out 30 cycles of amplification using a step programme as follows:
    - 94°C 1 min (denaturation)
    - 42°C 1 min (primer annealing step)
    - 72°C 3 min (extension step).

    The final extension step (cycle 30) should be for 15 min.

12. Analyse an aliquot of the PCR reaction by agarose gel electrophoresis. At this stage a smear of bands within the expected product size range is normal.

13. Dilute the amplification products from the first round of PCR 1:20 in TE. Amplify 1 μl of the diluted material using 25 pmol of each of the oligo(dT)$_{20}$ primer and the second internal sequence primer, P2 (see *Figure 5*). The PCR amplification should be carried out as in step **11**.

14. 10% of the product generated can be analysed on an agarose gel. The amplified product should now appear as a single band.

5.4.2 Amplification of sequences upstream of a known sequence

A primer which binds to a known sequence is used to initiate the synthesis of a cDNA strand (see *Figure 6*). This cDNA is then modified by the addition of a poly(A) tail. The efficiency of the cDNA extension in this reaction is crucially important. Each specific cDNA molecule, no matter how short, can become tailed. Thus the successful outcome of the PCR reaction is highly dependent on the reverse transcription reaction. The same 20mer oligo(dT) primer used in *Protocol 10* can be used in combination with two gene-specific primers to yield the desired product. The gene-internal primers utilized should be as close as possible to the 5′ end of the region of known sequence.

The addition of additional nucleotides to the 5′ end of the 20mer oligo(dT) primer (such as the recognition sites for restriction endonucleases) may improve the quality and yield of the amplified product. The addition of the extra bases to the primer allows the use of higher stringency conditions in the annealing step of the protocols. As the extra bases cannot hybridize to the

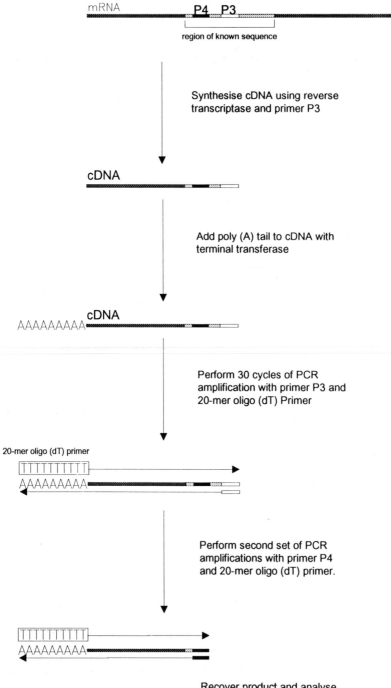

Figure 6. Amplification of the 5' ends of cDNAs by anchored PCR. The procedure is very similar to that shown in *Figure 5*. Initially a pool of RNA is reverse transcribed using a primer that binds to a region of known sequence within the gene of interest (P3 in the drawing). A synthetic poly(A) tail is then added to the cDNA pool with terminal transferase. The tailed cDNAs are then amplified using a 20mer oligo(dT) primer and the gene-specific primer, P3. The products generated are then re-amplified using a second gene-specific primer, P4, and the 20mer oligo(dT) primer. After re-amplification a single band should be visible on an ethidium bromide stained gel.

---

poly(A) tail the first two PCR cycles must be carried out under low stringency conditions. However, they will become incorporated during the synthesis of the complementary strand allowing the use of higher stringency conditions in later amplification cycles. If sites for restriction endonuclease are added to the 5' end of the oligo(dT) primer this may also facilitate the cloning of the amplified products.

Protocol 11. Amplification of regions upstream of the 5' end of a cDNA molecule

Equipment and reagents

- Automated thermal cycler
- Centricon-100 spin filters (Amicon Corp.)
- Speed-Vac desiccator (Savant)
- Source of RNA
- Moloney murine leukaemia virus (M-MLV) reverse transcriptase (200 U/μl)
- 2.5 U/μl *Taq* DNA polymerase (Boehringer-Mannheim or Perkin-Elmer Cetus)
- 25 pmol/μl each of sequence specific primers (P3 and P4)

- 20 pmol/μl oligo(dT)$_{20}$ primer
- 10 × amplification buffer (*Protocol 10*)
- 5 mM dNTP mix (*Protocol 10*)
- 5 × terminal transferase buffer: 125 mM Tris–HCl pH 6.6, 1 M potassium cacodylate, 1.25 mg/ml bovine serum albumin
- RNasin (Promega)
- Mineral oil
- TE (*Protocol 10*)

Method

1. Follow *Protocol 10*, steps 1–4. The 20mer oligo(dT) primer used in *Protocol 10* is replaced with a gene-specific primer (see *Figure 6*) to prime cDNA synthesis.

2. Dilute the reverse transcription reaction to 2 ml with TE. Excess primers are removed from the reaction mixture through the use of Centricon-100 spin filters (Amicon). Apply the reverse transcription mixture to the filter and spin for 20 min at 1000 *g*. Repeat the centrifugation using 2 ml of 0.2 × TE and collect the retained liquid.[a]

3. Lyophilize the sample in a Speed-Vac desiccator. Resuspend the pellet in 10 μl of H$_2$O and heat to 90°C for 3 min. Spin briefly in a microcentrifuge and place the sample on ice.

Protocol 11. *Continued*

4. Prepare the terminal transferase mix on ice:
- 4 µl of 5 × terminal transferase buffer
- 1.2 µl of 25 mM $CoCl_2$
- 4 µl of 1 mM dATP
- 0.8 µl (10 U) of terminal transferase.

Add the terminal transferase mix to the cDNA products from step **3**. Incubate at 37 °C for 10 min.

5. Inactivate the enzyme by heating for 5 min at 65 °C. Dilute to 0.5 ml with TE.

6. Perform the PCR amplification as in *Protocol 10*, steps 6–12. Use 1 µl of the tailed cDNA from step **5** as the template for the PCR reaction. The gene-specific primer used should be equivalent to P3 (see *Figure 6*). The same 20mer oligo(dT) primer as used in *Protocol 10* may be used.

7. Follow *Protocol 10*, steps 13 and 14. The diluted material should be amplified with the second gene-specific primer (P4 in *Figure 6*) and the 20mer oligo(dT) primer.

The PCR products yielded by *Protocols 10* and *11* should be monitored by Southern hybridization. The products of the first set of PCR reactions may be monitored by using a target specific internal oligomer (such as primers P2 and P4, see *Figures 5* and *6*). This will confirm that the desired target is being amplified in the early stages of the protocol. After the second set of PCR reactions almost 100% of the cDNA detected on an ethidium bromide stained gel should represent specific product. The products of the RACE-PCR reactions may be sequenced directly on a population level from the end at which the gene-specific primers are located (see Chapter 7), or the reaction products may be cloned into a suitable vector.

The PCR protocols detailed above can be used to obtain information on the use of alternate promoters, splicing, and polyadenylation of a given gene transcript. Additionally, any differentially spliced or initiated transcripts may also be visible after electrophoresis and cloned separately.

6. Analysis of protein–DNA interactions

Transcription in eukaryotes is controlled by nuclear proteins binding to specific sequence elements (21, 22). The assays described in this section are used to visualize the sequence-specific binding of purified preparations or crude mixtures of proteins to specific DNA sequences. We shall describe a set of complementary techniques, each with its specific advantages and shortcomings.

6.1 Preparation of crude nuclear extracts

A good nuclear extract is the starting point for many assays. It can be used for *in vitro* transcription, splicing, a source of DNA binding proteins, and as a starting point for biochemical fractionation. *Protocol 12* outlines the preparation of nuclear extracts from cells or tissues (23). In this procedure cells are allowed to swell in a hypotonic buffer, nuclei are then freed by homogenization and collected by centrifugation. Nuclear proteins are then salt extracted (400 mM NaCl or KCl). The majority of the chromatin complex will then be left as an insoluble residue. The nuclear extract is then dialysed. All operations should be performed on ice in a cold room. Centrifugation should be carried out at 4 °C using pre-cooled rotors.

A cytoplasmic extract may be made as a by-product (see *Protocol 12*, footnote *c*). Some nuclear proteins rapidly leak out of the nucleus after homogenization. These proteins can be found in the cytoplasmic extract which should therefore be saved.

Protocol 12. Preparation of nuclear extracts from cells or tissues

Equipment and reagents

- Protease inhibitors[a]
- Glass Dounce homogenizer with a type B pestle (the volume should be appropriate to the size of the extract being made, usually within the range 10–50 ml)
- Buffer A: 10 mM Hepes–KOH pH 7.9, 1.5 mM $MgCl_2$, 10 mM KCl, 0.5 mM dithiothreitol
- Buffer B: 0.3 M Hepes–KOH pH 7.9, 1.4 M KCl, 30 mM $MgCl_2$

- Buffer C: 20 mM Hepes–KOH pH 7.9, 25% glycerol, 400 mM NaCl, 1.5 mM $MgCl_2$, 0.2 mM EDTA, 0.5 mM dithiothreitol
- Buffer D: 20 mM Hepes–KOH pH 7.9, 20% glycerol, 100 mM KCl, 0.2 mM EDTA, 0.5 mM dithiothreitol
- Phosphate-buffered saline (PBS)
- Bio-Rad microassay kit (or equivalent) to measure protein concentration

Method

1. Collect cells grown in suspension by centrifugation.[b] Measure the packed cell volume (PCV). Wash by gently resuspending in 5 × PCV of PBS. Re-centrifuge and re-measure the PCV.

2. Resuspend the cells in five PCVs of buffer A.[a] Allow cells to swell by standing on ice for 10 min.

3. Homogenize by five to ten up and down strokes of a glass Dounce homogenizer (type B pestle). Centrifuge for 10 min at 1000 *g*.

4. Pour off the supernatant carefully.[c] Resuspend the nuclear pellet in 2.5 ml/10^9 cells of buffer C.

5. Homogenize as in step **3**.

6. Stir the nuclear suspension for 30 min and then centrifuge for 30 min at 25 000 *g*.

179

Protocol 12. *Continued*

7. Dialyse the supernatant for 4–5 h against > 50 PCVs of buffer D. The dialysed extract should then be centrifuged at 25000 *g* for 10 min. Freeze the supernatant in aliquots at −70 °C. The protein concentration should be determined (e.g. by using the Bio-Rad microassay). It should be in the range 5–10 mg/ml.

[a] The following protease inhibitors should be added to all buffers just before use: 0.25 mM phenylmethylsulfonyl fluoride (PMSF) and 0.25 mM benzamidine (Sigma). PMSF should be dissolved in propan-2-ol as a 250 mM stock solution. It is not advisable to store the PMSF stock (at −20 °C) for more than a few days. Other protease inhibitors may also be used.
[b] To isolate proteins from tissues it is necessary to cut the tissue into small pieces or briefly homogenize using a blender with rotating blades. The tissue is then washed two or three times with PBS.
[c] This supernatant may be used to make a cytoplasmic extract. Add 0.11 vol. (as a fraction of the supernatant volume) of buffer B to the supernatant. Centrifuge for 60 min at $1 \times 10^6 g$. Dialyse the supernatant 5–8 h against > 20 vol. of buffer D (see step 7). Freeze in aliquots at −70 °C.

6.2 DNA binding assays

In each of the DNA binding assays described below the optimal conditions must be determined empirically. The following may all affect the stability of a DNA–protein complex:

(a) The monovalent cation concentration. Proteins generally bind more strongly to DNA at low K^+ or Na^+ concentrations.

(b) The divalent cation concentration. Mg^{2+}, Ca^{2+}, and Zn^{2+} ions have differing effects on the stability of various DNA–protein complexes. Altering the Mg^{2+} concentration usually has the biggest effect. Zn^{2+} at 1 μM may be added if studying the binding of a putative zinc finger protein to DNA.

(c) The addition of competitor DNA. Nuclear extracts contain a large number of DNA binding proteins which will bind non-specifically to the labelled DNA fragment. The choice of competitor DNA is crucial to the success of the experiment. Poly(dI-dC) : poly(dI-dC) is the most commonly used competitor. Poly(dI-dC) : poly(dI-dC) is usually most effective when working with fractionated nuclear extracts. When a crude nuclear extract is used, high concentrations of poly(dI-dC) : poly(dI-dC) alone may not be sufficient to bind all non-specific DNA binding proteins. In such cases poly(dI-dC) : poly(dI-dC) may be used in combination with other competitor DNAs. These include linearized plasmids such as pBR322 and natural DNAs isolated from *E. coli* and salmon sperm. The competitor DNA sequence should differ as much as possible from that of the anticipated protein binding site.

(d) Volume excluders such as polyvinyl alcohol or polyethylene glycol. By bringing about an increase in the effective protein and DNA concentration these agents have been observed to stimulate the binding of purified proteins to DNA.

6.2.1 Gel retardation assay

In this assay a labelled DNA fragment is incubated with an excess of a protein sample and the mixture analysed on a non-denaturing polyacrylamide gel. The formation of a protein complex on the probe DNA is reflected in the decreased mobility of this complex in native gels compared to that of the free probe (24). Bandshifting is a fast and sensitive assay (once the conditions have been established), and is ideal for the screening of column fractions during protein purification. The end-labelled probes for bandshifting should be as small as possible (ideally 30–100 bp long) in order to minimize the binding of non-specific proteins. Either synthetic oligonucleotide probes or short restriction fragments containing the protein binding site(s) of interest may be used.

Crude nuclear extracts contain a large number of proteins that can bind non-specifically to DNA. In order to minimize the effects of this, it is necessary to include an excess of non-specific competitor DNA in the assay (see above). If substantial amounts of radioactivity do not enter the gel the concentration of poly(dI-dC) : poly(dI-dC) competitor should be increased. If this has little effect an additional competitor (e.g. linearized plasmid DNA) may be used.

The gel retardation assay can be prone to artefacts. Protein–DNA complexes observed by this method have to be carefully examined to verify whether they are caused by a specific interaction at the site of interest. This can be done by competition with an excess of specific and non-specific DNA (see *Figure 7*). The use of mutant probes that differ from the wild-type by point mutations can also help to determine whether the protein–DNA complex observed is the result of a specific interaction. The exact site at which a protein is interacting with the probe may be determined by combining the bandshift and footprinting assays (see below).

Protocol 13. Bandshift assay

Equipment and reagents

- Vacuum gel drier
- Non-denaturing polyacrylamide gel (prepared as in Chapter 4, *Protocol 9*)
- 10 × binding buffer: 500 mM KCl, 200 mM Tris–HCl pH 7.6, 50 mM MgCl$_2$, 10 mM EDTA, 4 mM spermidine, 2 mM DTT (store at −20 °C)
- 40% Ficoll
- 3 mg/ml poly(dI-dC) : poly(dI-dC)[a]
- End-labelled DNA fragment or double-stranded oligonucleotide containing the protein binding site at 10 000 c.p.m./μl (see Chapter 4)
- Protein extract (*Protocol 12*)

Protocol 13. *Continued*

Method

1. Mix the following reagents in a final volume of 20 μl:
   - 2 μl 10 × binding buffer
   - 4 μl 40% Ficoll
   - 2 μl poly(dl-dC) : poly(dl-dC)[a]
   - 1 μl ^{32}P end-labelled DNA fragment
   - 2 μl protein extract[b]
   - H₂O to a final volume of 20 μl.[c]

2. Incubate the mixture on ice for 15 min.

3. Load into one well of the non-denaturing polyacrylamide gel and electrophorese at 10–15 V/cm. The gel must not warm up. Bromophenol blue and xylene cyanol dyes may be loaded in an adjacent lane as markers. For a 40mer oligonucleotide probe the electrophoresis takes about 90 min.

4. The gel may then be dried down after transfer to paper using a vacuum gel drier. An autoradiograph is then made from the dried gel at −70 °C (usually an overnight exposure is required).

[a] Poly(dl-dC) : poly(dl-dC) is dissolved in TEN (10 mM Tris–HCl pH 7.6, 1 mM EDTA, 50 mM NaCl). Poly(dl-dC) : poly(dl-dC) may not be necessary if working with affinity purified protein.
[b] 5 μg of a crude nuclear extract is usually sufficient to detect ubiquitous DNA binding proteins in this assay. The volume of protein used in this assay should be altered accordingly. If higher concentrations of protein are used the amount of competitor DNA added should be increased in proportion.
[c] If performing a bandshift competition experiment unlabelled specific competitor DNAs may be included in the incubation mix (see *Figure 7*)

6.2.2 Methylation interference

This method can be used to study the specificity of a protein–DNA interaction observed in the gel retardation assay. The end-labelled DNA fragment is partially methylated with dimethyl sulfate (DMS) which methylates guanine at N7 (major groove) and to a lesser extent adenine at N3 (minor groove). This probe is then incubated with a protein extract and the bound and unbound DNA separated in a bandshift reaction. Probe molecules that carry methyl groups in a position that are essential for protein binding are selectively missing from the retarded fraction.

In order to obtain sufficient radioactive counts in the retarded band the bandshift assay detailed in *Protocol 13* is scaled up. Around 4×10^5 c.p.m. of end-labelled probe are incubated with 40 μl of protein extract. A sufficient number of counts must be used in order to detect the retarded band when the

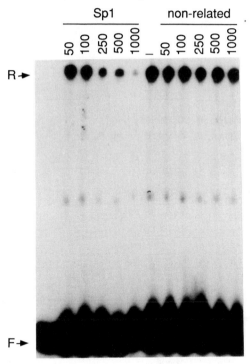

Figure 7. Gel retardation assay. In the experiment shown the binding of transcription factor Sp1 to a double-stranded 30mer oligonucleotide probe is studied. Competition with increasing molar excesses of unlabelled probe oligonucleotide (50, 100, 250, 500, and 1000 times) leads to decreasing amounts of protein in the retarded complex (R). In contrast, increasing amounts of a competitor oligonucleotide containing an unrelated sequence has no effect on the specific binding of Sp1 to the probe. No competitor DNA has been added to the bandshift reaction in the *centre* of the gel (labelled −). From the experiment shown it can be deduced that Sp1 specifically recognizes its binding site on the oligonucleotide probe and has little or no affinity for other DNA sequences. Unbound probe is labelled F.

gel is autoradiographed at room temperature for one to two hours. The DNA is then extracted from the retarded and non-retarded bands and cleaved with piperidine. The cleavage products are then analysed on a sequencing gel (see *Figure 8*).

The DNA that is run on the sequencing gel must be free from salts, piperidine, and acrylamide contaminants. If working with DNA fragments larger than 80 bp the best way of cleaning up the DNA appears to be Elutip D columns (Schleicher and Schuell). In most cases the method described in *Protocol 14* works well. If the piperidine cleavage does not go to completion the bands will be very fuzzy.

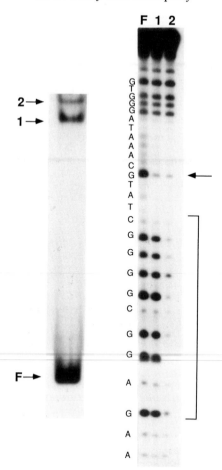

Figure 8. Methylation interference. The *left* panel shows a bandshift reaction with a 40mer oligonucleotide probe containing the adjacent octamer and Sp1 binding sites in the *Xenopus* U2 gene distal sequence element (DSE). The bandshift reaction was set-up as in *Protocol 14* except that 10% glycerol (v/v) was included in the reaction buffer in place of 8% Ficoll. The retarded complexes 1 and 2 were extracted from the gel together with the DNA contained in unbound probe fraction, F. After piperidine cleavage the reaction products were run on a 15 % denaturing polyacrylamide gel as shown in the *right* panel. Part of the oligonucleotide sequence is written down the left-hand side. The retarded complex 1 arises through proteins binding to the octamer motif (TAAACGTA as written from top to bottom in the sequence lane). The octamer protein only binds efficiently to its binding site if it is completely unmethylated. The central G in the octamer binding site is *arrowed*. In contrast, retarded complex 2 arises through both the octamer and Sp1 factors binding to the U2 DSE probe. The Sp1 site, CGGGGCGGAG, is bracketed. All of the G residues in this region are not methylated as can be seen by comparing lanes F and 2.

Protocol 14. Methylation interference

Equipment and reagents

- Geiger counter
- Speed-Vac centrifugal desiccator
- Fume hood
- End-labelled double-stranded oligonucleotide or DNA fragment (see Chapter 4, *Protocols 4–8*)
- Non-denaturing polyacrylamide gel (prepared as in Chapter 4, *Protocol 8*)
- Sequencing gel (prepared as in Chapter 7, *Protocol 8*)
- G buffer: 50 mM sodium cacodylate, 0.1 mM EDTA pH 8.0

- G-stop buffer: 1.5 M sodium acetate, 1.0 M Tris–acetate pH 7.5, 1.0 M β-mercaptoethanol, 1 mM EDTA
- Dimethyl sulfate (DMS) (reagent grade)
- Protein extract (*Protocol 12*)
- poly(dI-dC) : poly(dI-dC) (3 mg/ml)
- 10 × binding buffer (*Protocol 13*)
- 40% Ficoll (type 400)
- Piperidine
- Formamide dyes (*Protocol 7*)

Method

Caution! Methylation of the end-labelled probe must be carried out in a fume hood as DMS is a carcinogen.

1. End-label and purify the oligonucleotide or restriction fragment as described in Chapter 4, *Protocols 4–8*.

2. Add 200 µl of G buffer and 0.5 µl of DMS (pre-mixed) to $> 400\,000$ c.p.m. of probe (in a minimum volume). Mix and incubate on ice for 1 min.

3. Add 50 µl of G-stop buffer and 750 µl of cold ethanol. Mix well and freeze in a dry ice/ethanol bath.

4. Centrifuge for 30 min to recover the DNA. Check with a Geiger counter that most of the counts have been precipitated. Wash with 70% ethanol. Dry and take up in 10 µl H_2O.

5. Set up a bandshift reaction as follows:[a]
   - 4×10^5 c.p.m. methylated probe
   - 5 µl nuclear extract (about 40 µg of protein)
   - 7 µl poly(dI-dC) : poly(dI-dC) (3 mg/ml)
   - 4 µl 10 × binding buffer
   - 8 µl 40% Ficoll
   - Make up to 40 µl with H_2O.

6. Incubate on ice for 15 min.

7. Load on to a 6% non-denaturing gel. The gel used should be 1 mm thick and the comb should create a slot 15 mm wide. Carry out electrophoresis at 12 V/cm. Bromophenol blue and xylene cyanol dyes should be run in an adjacent lane as markers.

8. After electrophoresis remove one glass plate from the gel. Cover the

185

Protocol 14. *Continued*

gel attached to the other plate with cling-film and place gel on X-ray film in a cassette. Mark the position of the gel plate. Expose the gel for 1–2 h. The amount of time needed can be determined by running a Geiger counter over the gel. The position of the retarded band should be easily detectable.

9. Cut out all bands and elute the DNA as in Chapter 4, *Protocol 8*.

10. Re-dissolve DNA in 100 μl of 10% (1 M) piperidine solution (freshly diluted). Incubate at 90 °C for 30 min.

11. Cool on ice. Spin down briefly in a microcentrifuge. Lyophilize for at least 2 h or overnight using a Speed-Vac centrifugal desiccator.

12. Re-dissolve the DNA in 40 μl of H_2O and lyophilize for 60 min.

13. Re-dissolve the DNA in 40 μl of ddH_2O. Spin down briefly. Transfer the supernatant containing soluble counts only to a new tube. Lyophilize for 60 min.

14. Take up DNA in 10 μl of formamide dyes. Heat at 90 °C for 3 min and put on ice. Load around 5×10^3 c.p.m. per slot on to a sequencing gel.

[a] The conditions given for the binding reaction above are the same as those for the experiment shown in *Figure 8*. For a given probe–DNA complex the binding conditions will have to be determined empirically to obtain sufficient counts in the retarded band(s). In many cases the poly(dI-dC):poly(dI-dC) competitor alone will not be sufficient and plasmid DNA will probably have to be included in the binding reaction.

6.2.3 DNase I footprinting

DNA fragments containing the binding site for the protein under study are radioactively labelled at one end. Bound protein protects the probe from enzymatic attack by DNase I which would otherwise lead to DNA cleavage. Comparison of the DNase I cutting pattern in the presence and absence of a protein (or proteins) yields information about the protein binding site (see *Figure 9*) (25). The protein binding site(s) can be localized within a few nucleotides.

Footprinting lacks the sensitivity of the gel retardation assay (*Protocol 13*). The method requires high occupancy (50–100%) of the binding site(s) on the probe to give interpretable signals. It is sometimes difficult to obtain good footprints with crude nuclear extracts and some fractionation of the extract (e.g. by passing it over a heparin–Sepharose column) may be necessary.

Footprinting probes are usually restriction fragments labelled as in Chapter 4, *Protocols 5–7*. Ideally the site of interest (if known) should be at least 50 bp from the labelled end (to minimize interference from non-specific 'end-binding' proteins) and at most 150 bp (to allow good resolution on sequencing gels). If the protein binding site is too close to the unlabelled end, non-specific end-binding can again be problematic.

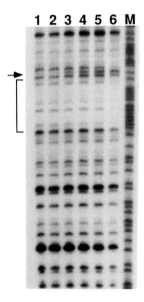

Figure 9. DNase I footprinting. The figure shows binding of affinity purified Sp1 to its recognition sequence, CGGGGCGGAG, in the *Xenopus* U2 gene distal sequence element (DSE). Lane 1, no protein added; Lanes 2–6, 0.25, 0.5, 1.0, 2.0, and 4.0 μl of the affinity purified protein; lane M, a Maxam and Gilbert G + A sequencing reaction performed on the end-labelled probe. The binding of the purified protein can be first detected by the appearance of a new DNase I band (*arrowed*). The footprint over the Sp1 site (bracketed) can only be clearly seen in lane 6.

It is possible to examine the contacts a protein makes with both strands of the DNA molecule by performing 5′ and 3′ end-labelling. For example the 5′ overhang of a restriction fragment may be labelled with polynucleotide kinase. A complementary 3′ end-labelled probe can be generated by filling in the 5′ overhang with the Klenow or reverse transcriptase enzymes.

Protocol 15. DNase I footprinting

Equipment and reagents

- End-labelled probe prepared as in Chapter 4, *Protocols 5–7*
- Polyvinyl alcohol (10% w/v)
- Linearized plasmid DNA (12.5 μg/ml)
- Poly(dl-dC):poly(dl-dC) (3 mg/ml)
- Footprinting buffer: 25 mM Hepes–KOH pH 7.8, 1 mM dithiothreitol, 20% (v/v) glycerol, 0.1% (v/v) Nonidet P-40, 100 mM KCl
- 10 mM $MgCl_2$, 5 mM $CaCl_2$
- DNase I (Cooper Biomedical, Worthington DPFF; make a 3 mg/ml stock solution)

- Stop buffer: 20 mM EDTA pH 8.0, 1% SDS, 200 mM NaCl, 250 μg/ml carrier RNA
- Phenol/chloroform (25 : 24 : 1 phenol/chloroform/isoamyl alcohol)
- Formamide dyes (*Protocol 7*)
- 6% sequencing gel (see Chapter 7)
- Maxam and Gilbert G + A cleavage reaction on the probe (Chapter 4, *Protocol 9*)

Protocol 15. *Continued*

Method

1. Make up a probe DNA mix (for each reaction):
   - 10 000 c.p.m. ^{32}P end-labelled DNA fragment
   - 10 µl of 10% (w/v) polyvinyl alcohol
   - 1 µl of 12.5 µg/ml plasmid DNA (e.g. pUC)[a]
   - 0.5 µl of 3 mg/ml poly(dl-dC):poly(dl-dC)[a]
   - water to a final volume of 25 µl.

2. Adjust the protein fractions to 25 µl in footprinting buffer, which is similar to buffer D (*Protocol 12*).

3. Combine the probe mix and protein mix. Incubate on ice for 15 min.

4. Add 50 µl of 10 mM $MgCl_2$, 5 mM $CaCl_2$. Let stand at room temperature for 1 min.

5. Add 2 µl of DNase I at the appropriate dilution[b] to each sample. Mix and incubate at room temperature for exactly 1 min.

6. Stop the reaction by the rapid addition of 90 µl of stop buffer.

7. Add 200 µl of phenol/chloroform. Mix by vortexing. Centrifuge for 3 min to separate the phases. Transfer upper phase to a new test-tube.

8. Add 750 µl of ethanol. Mix and centrifuge for 15 min. Wash the pellet with 70% ethanol and dry briefly.

9. Re-dissolve the pellet in 5 µl formamide dyes. Heat to 90 °C for 3 min and transfer to ice. Load 2.5 µl of each sample on a sequencing gel (see Chapter 7) together with the products of a Maxam and Gilbert G + A cleavage reaction on the probe to provide precise size markers (see Chapter 4, *Protocol 9*).

---

[a] The concentration of competitor DNAs must be determined empirically. If using affinity purified proteins the addition of competitor DNAs is usually not necessary. The conditions given are for footprinting Sp1 from a crude nuclear extract of HeLa cells.

[b] The DNase I concentration required must be determined empirically. A 1/2000 dilution from a 3 mg/ml stock usually works well for the no protein control lanes. For footprinting Sp1 from 20 µg of HeLa nuclear extract a 1/300 dilution worked well. In general the concentration of DNase I used must be increased in proportion to the amount of protein in the footprinting reaction. Some nuclear extracts contain inhibitors of DNase I activity.

6.2.4 Other footprinting methods

i. Exonuclease protection

This method is much more sensitive than DNase I footprinting. 5' End-labelled DNA is incubated with protein. Exonuclease III is then used to digest each strand from the 3' end until it encounters the protein specifically bound to the DNA (26) (*Figure 10*). The length of the undigested labelled

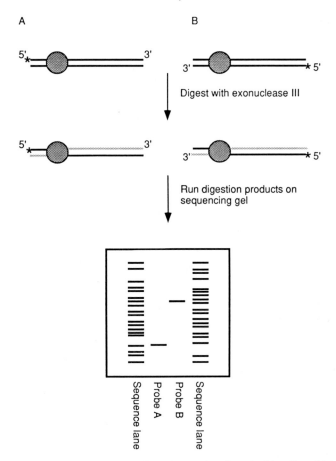

Figure 10. Exonuclease III footprinting. Proteins are incubated with a 5′ end-labelled DNA probe. In panel A the restriction fragment is labelled at the left-hand end and in panel B the restriction fragment is labelled at the right-hand end. The DNA is then digested with exonuclease III. A protein bound to the DNA halts the 3′–5′ progress of the enzyme along the DNA strand. In panel A a short DNA fragment is protected from exonuclease III digestion, whereas in panel B a somewhat longer DNA fragment is protected. The reaction products are run on a denaturing polyacrylamide gel together with Maxam and Gilbert sequencing reactions performed on the end-labelled probe. The length of the DNA fragments protected from exonuclease III digestion enable the protein binding site to be mapped.

DNA then reveals the position of one of the boundaries of the protein binding site. The other boundary can be determined by labelling the restriction fragment at the opposite end and performing an exonuclease III digestion reaction. Despite its sensitivity the exonuclease III footprinting assay is not widely used. Problems can arise in assigning the exonuclease III stops to positions of protein binding to DNA as opposed to secondary structure effects, non-specific protein binding, or multiple proteins binding to the DNA.

ii. Chemical footprinting reagents

A range of chemical footprinting reagents have been described in the literature over the past ten years (27–29). Methyl propidium EDTA (MPE), an intercalating agent, is the most widely used and cleaves DNA by radical attack (27). Prevention of MPE intercalation by protein binding leads to the formation of the footprint. MPE reacts at a similar rate with each nucleotide of the probe. MPE footprints are easily distinguished by the very uniform size distribution of the reaction products. The chemistry of the reaction and the fact that MPE is a much smaller molecule than DNase I allows the binding site of a protein to DNA to be very closely defined. Several factors have been found to inhibit MPE activity, including high levels of divalent cations or reducing agents. The presence of glycerol in the reaction buffer is especially inhibitory. Furthermore it is necessary to achieve virtually 100% occupancy of the protein binding site on the probe before a good footprint can be observed.

7. Concluding remarks

In this chapter we have collected together a number of core techniques that we consider central to the screening for cloned genes and the study of their organization and transcription. To provide a comprehensive survey of all the methods used in such analyses would involve the production of a huge volume in itself and the reader is referred to the excellent laboratory manuals by Sambrook *et al.* (2) and Ausubel *et al.* (30). An excellent collection of protocols for the analysis of gene transcription has also been presented in another volume in this series (31). Nevertheless the methods presented form a useful core for preliminary analyses of gene expression. Once the regulatory regions of genes have been mapped by mutagenesis and the detection of DNA binding proteins a start can be made on the functional analysis of gene expression. It is now very straightforward to purify transcription factors and other DNA binding proteins by chromatography. Purified proteins or protein fractions can then be used to build up a picture of the interactions involved in controlling gene expression in *in vitro* transcription experiments.

References

1. Kadonaga, J. T. and Tjian, R. (1986). *Proc. Natl. Acad. Sci. USA.*, **83**, 5889.
2. Sambrook, J., Fritsch, E. F., and Maniatis, T. (ed.) (1989). *Molecular cloning: a laboratory manual* (2nd edn). Cold Spring Harbor Laboratory Press, Cold Spring Harbor, NY.
3. Schwartz, D. C. and Cantor, C. R. (1984). *Cell*, **37**, 67.

4. Anand, R. and Southern, E. M. (1990). In *Gel electrophoresis: a practical approach* (ed. D. Rickwood and B. D. Hames), pp. 101–22. IRL Press, Oxford.
5. Southern, E. M. (1975). *J. Mol. Biol.*, **98**, 503.
6. Smith, G. E. and Summers, M. D. (1980). *Anal. Biochem.*, **109**, 123.
7. Alwine, J. C., Kemps, D. J., and Stark, G. R. (1977). *Proc. Natl. Acad. Sci. USA*, **74**, 5350.
8. Stallcup, M. R. and Washington, L. D. (1983). *J. Biol. Chem.*, **258**, 2802.
9. Church, G. M. and Gilbert, W. (1984). *Proc. Natl. Acad. Sci. USA*, **81**, 1991.
10. Bolton, E. T. and McCarthy, D. J. (1962). *Proc. Natl. Acad. Sci. USA*, **48**, 1390.
11. Itakura, K., Rossi, J. J., and Wallace, R. B. (1984). *Annu. Rev. Biochem.*, **53**, 323.
12. *The DIG system user's guide for filter hybridisation.* (1993). Boehringer-Mannheim, GmbH Biochemica, D-6800 Mannheim 31, Germany.
13. McPherson, M. J., Quirke, P., and Taylor, G. R. (ed.) (1991). *PCR: a practical approach*. IRL Press, Oxford.
14. Gonzalez, C. and Glover, D. M. (1994). In *The cell cycle: a practical approach* (ed. P. Fantes and R. Brooks), pp. 143–75. IRL Press, Oxford.
15. *Non-radioactive* in situ *hybridisation—application manual*. (1992). Boehringer-Mannheim, Germany.
16. Favaloro, J. R., Treisman, R., and Kamen, R. (1980). In *Methods in enzymology* (ed. L. Grossman and K. Moldave), Vol. 65, pp. 718–29. Academic Press.
17. Zinn, K., DiMaio, D., and Maniatis, T. (1983). *Cell*, **34**, 865.
18. Ochman, H., Ayala, F. J., and Hartl, D. L. (1993). In *Methods in enzymology* (ed. R. Wu), Vol. 218, pp. 309–21. Academic Press.
19. Frohman, M. A., Dush, M. K., and Martin, G. R. (1988). *Proc. Natl. Acad. Sci. USA*, **85**, 8998.
20. Frohman, M. A. (1993). In *Methods in enzymology* (ed. R. Wu), Vol. 218, pp. 340–56. Academic Press.
21. Mitchell, P. J. and Tjian, R. (1986). *Science*, **245**, 371.
22. Serfling, E., Jasin, M., and Schaffner, W. (1985). *Trends Genet.*, **1**, 224.
23. Dignam, J. D., Leibowitz, R. M., and Roeder, R. G. (1983). *Nucleic Acids Res.*, **11**, 1475.
24. Fried, M. and Crothers, D. M. (1981). *Nucleic Acids Res.*, **9**, 6505.
25. Galas, D. J. and Schmitz, A. (1978). *Nucleic Acids Res.*, **5**, 3157.
26. Wu, C. (1985). *Nature*, **317**, 84.
27. van Dyke, M. W. and Dervan, P. B. (1983). *Nucleic Acids Res.*, **11**, 5555.
28. Spassky, A. and Sigman, D. S. (1985). *Biochemistry*, **24**, 8050.
29. Tullius, T. D. and Dombroski, B. A. (1986). *Proc. Natl. Acad. Sci. USA*, **83**, 5469.
30. Ausubel, F. M., Brent, R., Kingston, R. E., Moore, D. D., Seidman, J. G., Smith, J. A., and Struhl, K. (1993, updated quarterly). *Current protocols in molecular biology*. Greene Publishing Associates and Wiley-Interscience, 605 Third Avenue, New York.
31. Hames, B. D. and Higgins, S. J. (ed.) (1993). *Gene transcription: a practical approach*. IRL Press, Oxford.

6

Oligonucleotide-directed mutagenesis with single-stranded cloning vectors

HARALD KOLMAR and HANS-JOACHIM FRITZ

1. Introduction and overview

1.1 Basic principles

Directed mutagenesis with synthetic oligonucleotides was proposed in 1971 by Hutchison and Edgell (1) and first developed to a practical procedure by Zoller and Smith (2, 3). According to the original protocol, the gene of interest is first cloned in a vector derived from phage M13 (single-stranded circular DNA genome). An oligonucleotide is synthesized that is complementary to the target site of mutagenesis with the exception of one or more non-complementary nucleotide central residues. The oligonucleotide is annealed to the single-stranded, recombinant vector DNA, and covalently closed double-stranded heteroduplex DNA (hd DNA) containing a single site of non-homology is synthesized in an enzyme catalysed reaction. This hd DNA is transfected into *E. coli* where it replicates *in vivo*, to segregate DNA clones carrying the predetermined mutation carried on the synthetic oligonucleotide.

Some of the first spectacular successes of protein engineering were achieved with this method (see for example ref. 4). It was soon realized, however, that the procedure is bound to produce only low yields of the desired mutants because cellular processes of DNA mismatch repair are directed against the DNA strand synthesized *in vitro* (5). Marker yield is indeed an important aspect, since it drastically influences the work needed to identify a clone with the correct structure, which—in the absence of a phenotype—has to be accomplished solely by biochemical means such as nucleotide sequence analysis (5). As a consequence, a number of improved methods were devised all of which address the problem of marker loss associated with mismatch repair (for review see ref. 6). These improved methods can be grouped in different categories, depending on whether they:

- avoid the problem by enzymatic conversion of the hd DNA to fully complementary DNA already *in vitro*

- genetically abolish DNA mismatch repair
- recruit cellular DNA repair processes to work in favour of the synthetic marker.

In this chapter we give a brief outline of four commonly used second or third generation methods of oligonucleotide-directed mutagenesis based on single-stranded cloning vectors, describing three of them in experimental detail. The application of strong biochemical or genetic selection against the original ('wild-type') DNA template is a common feature of the first three methods described in this chapter.

1.2 The phosphorothioate method

In the phosphorothioate method of directed mutagenesis (7, 8), preferential establishment of the synthetic marker is achieved by strand-selective repair of hd DNA *in vitro* with concomitant physical elimination of the wild-type sequence. By the time the transforming DNA enters the cell, it is already in a homoduplex state as a result of involved enzymatic DNA processing *in vitro*. Thus, cellular DNA mismatch repair is irrelevant.

The mutagenic oligonucleotide serves as a primer for an *in vitro* DNA polymerase reaction in which, for example, dCTP is replaced by 2′-deoxy-cytidine-5′-O-(α-thiotriphosphate) (dCTPαS). This results in the generation of internucleotide phosphorothioate bridges in the complementary strand (*Figure 1*). The template strand, but not the newly synthesized strand, is then nicked by a restriction endonuclease, such as *Nci*I, which is unable to cleave a target site with an internucleotide phosphorothioate linkage. The nicked duplex DNA is subjected to limited exonucleolytic digestion with exonuclease III to produce a gap spanning the site of mutation. After re-polymerization, the resulting homoduplex DNA is used for transformation. This procedure has been described in explicit experimental detail in another issue of this series (7).

1.3 The gapped duplex DNA method

Both the gapped duplex DNA method (9, 10) and the deoxyuridine method exploit *E. coli* genetics to enrich the synthetic marker. The key intermediate in the gapped duplex approach is a partial DNA duplex, which has only the target region for mutation exposed in single-stranded form (*Figure 2*). The system has been devised such that the gapped duplex DNA (gd DNA) carries different and easily selectable genetic markers in the two DNA strands. In the pMa/c series of mutagenesis vectors (11), these are genes conferring resistance to the antibiotics ampicillin or chloramphenicol, respectively. In the course of an *in vitro* reaction to fill and seal the gaps, the mutagenic oligonucleotide becomes physically integrated into the shorter strand of the gd DNA and, as a result, the synthetic marker is genetically linked to the

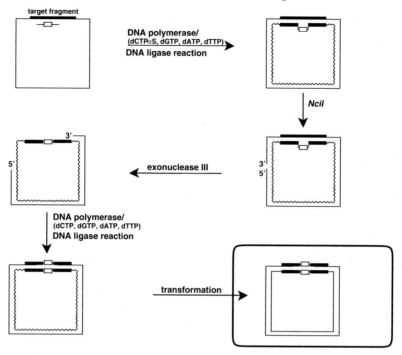

Figure 1. Directed mutagenesis using the phosphorothioate method. The presence of internucleotide phosphorothioate bridges is indicated by sinous lines. The various steps are discussed in the text.

particular antibiotic resistance marker carried by this DNA strand. Hence, selection for this antibiotic resistance after transformation and marker segregation co-selects for the synthetic marker. Marker scrambling is suppressed by use of a transfection host which is deficient in DNA mismatch repair. In successive multiple rounds of mutagenesis, selections for one or the other antibiotic resistance are applied in a reciprocating manner (11).

A gapped duplex DNA procedure applying the cellular *mutHLS* DNA mismatch repair pathway for marker enrichment (5) is only rarely used nowadays since its efficiency was found to depend on the nature of the base/base mismatch (12).

1.4 The deoxyuridine method

Marker enrichment in the deoxyuridine method (13–15) is achieved by recruiting the cellular DNA repair system responsible for removal of DNA uracil residues which, in the natural situation, arise through hydrolytic deamination of DNA cytosine residues. The uracil base is removed by uracil *N*-glycosylase, the product of the *ung* gene (16), and the resulting apyrimidinic site is

195

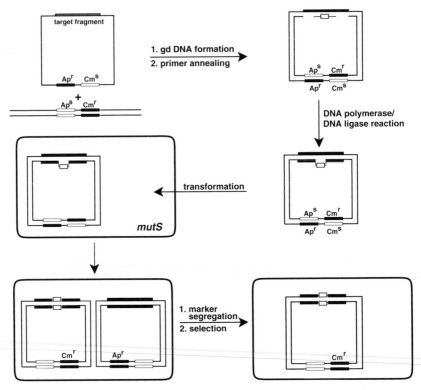

Figure 2. Schematic representation of the gapped duplex DNA method for oligonucleotide-directed mutation construction. Mutations can be introduced into a target fragment by cycling between pMa type (Ap^r, Cm^s; this contains an amber mutation in the Cm marker) and pMc type (Ap^s, Cm^r; this contains an amber mutation in the Ap marker). The wild-type antibiotic resistance genes are shown by filled rectangles, the amber mutant counterparts by open rectangles. For details refer to the text.

channelled into the general pathway of excision repair (17). The Ung enzyme acts on DNA uracil residues irrespective of whether they are matched, mismatched, or part of a single DNA strand.

For oligonucleotide-directed mutagenesis, single-stranded DNA to be mutagenized is grown in an *E. coli dut, ung* double mutant. Since the product of the *dut* gene (dUTPase) (18), is necessary to keep the cellular dUTP pool low, DNA grown in a *dut/ung* host contains a significant proportion of U instead of T (19). The mutagenic oligonucleotide (*Figure 3*) is annealed to this uracil-containing template and the complementary strand is synthesized by *in vitro* DNA polymerase/DNA ligase reaction using the four regular nucleotide triphosphates. The resulting heteroduplex DNA is used for transformation of an Ung$^+$ strain, with the result of repair initiation at every 2'-dU site and deactivation of genetic markers encoded by the uracil-containing DNA

196

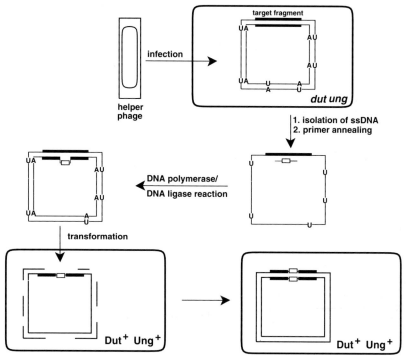

Figure 3. Directed mutagenesis using the deoxyuridine method. The various steps are discussed in the text.

strand. Since this strand contains the mutagenic oligonucleotide, the synthetic marker is strongly enriched.

1.5 The MHT protocol

In the mix-heat-transform (MHT) procedure (10, 20), recipient cells are transformed by a mixture of single-stranded DNA and mutagenic oligonucleotide without any enzymatic manipulations *in vitro*. Due to its simplicity, the method is particularly robust. Since, however, abolishing cellular DNA mismatch repair is the only genetic measure taken to counteract marker loss, yields of synthetic marker are generally low (typically a few per cent), which makes it advisable to combine the procedure with a generally applicable genetic screen for successful construction of the desired mutation. This is achieved by fusing the target gene, terminated by an amber stop codon, in frame to the phasmid-borne *lacZα* gene fragment (*Figures 4* and *5*), which is preceded by a polylinker. Expression of this fusion gene in an amber suppressor host leads to a blue colony phenotype on Xgal indicator plates. For introduction of the desired mutation, a single nucleotide within or near the

197

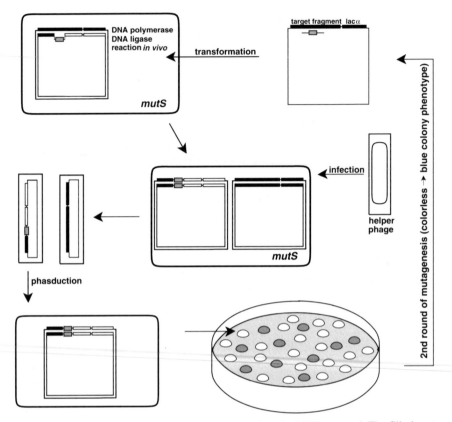

Figure 4. Consecutive introduction of mutations using the MHT protocol. The filled rectangles indicate expression of the target fragment/*lac*Zα fusion gene. Open rectangles indicate the presence of a frameshift mutation introduced by the mutagenic oligonucleotide. The various steps are discussed in the text.

codon of interest is removed in a first mutagenesis round. This frameshift results in a colourless colony phenotype. In a second round of mutation construction, the desired mutation is introduced together with restoration of the reading frame; hence successful mutation construction is indicated by a blue colony phenotype and marker yield does not play a major role.

2. Materials

2.1 Mutagenesis kits

Vectors, *E. coli* strains, buffers, and enzymes for the different mutagenesis protocols described in this chapter are available from the following commercial suppliers:

- phosphorothioate method: Amersham
- deoxyuridine method: Bio-Rad Laboratories; Boehringer-Mannheim Biochemicals.

Vectors and strains for the gapped duplex DNA and the MHT method are freely available from the authors upon request.

2.2 Enzymes and nucleoside triphosphates

The following list of reagents indicates suppliers used by the laboratory of the authors; enzymes and chemicals of equivalent quality may also be obtained from different suppliers.

- polynucleotide kinase: Boehringer-Mannheim Biochemicals
- T7 DNA polymerase form II unmodified: New England Biolabs
- T4 DNA ligase: Boehringer-Mannheim Biochemicals
- 2'-deoxynucleoside 5'-triphosphates (dATP, dGTP, dCTP, dTTP) and adenosine 5'-triphosphate (ATP), ultrapure 100 mM stock solutions: Pharmacia LKB Biotechnology.

2.3 Bacterial strains and helper phage

Both phasmid and helper phage require for infection the presence of pili on the surface of the host cells. Functional pili are formed by all strains noted in this chapter. Since pili are not formed at growth temperatures below 35 °C, all bacterial cultures should be incubated at 37 °C. Overnight cultures should be grown in glucose minimal medium. For BMH71-18/*mutS* tetracyclin at a concentration of 20 µg/ml should be added.

A method for the preparation of competent *E. coli* cells is described in *Protocol 1*. With this procedure, competent cells may be used when stored on ice for two to three days. Alternatively, the protocols described in Chapter 1 may be used.

Protocol 1. Preparation of *E. coli* competent cells and transformation

Reagents

- Overnight culture of host *E. coli* strain grown in glucose minimal medium—for growth of strain BMH71-18/*mutS*,[a] glucose minimal medium should be supplemented with tetracyclin at a final concentration of 20 µg/ml
- 2YT medium: 16 g tryptone, 10 g yeast extract, 5 g NaCl, made up with water to 1 litre and sterilized by autoclaving
- 2YT agar: 16 g tryptone, 10 g yeast extract, 5 g NaCl, 15 g Bacto agar made up with water to 1 litre and sterilized by autoclaving

- Glucose minimal medium: 7 g Na_2HPO_4. $2H_2O$, 3 g KH_2PO_4, 1 g NH_4Cl, make up with water to 1 litre, sterilize by autoclaving, and add the following sterile solutions: 25 ml 20% (w/v) glucose, 1 ml 100 mM $CaCl_2$, 1 ml 1 M $MgSO_4$, 5 ml 0.1 mM $FeCl_3$, 1 ml 0.1% (w/v) filter sterilized thiamin
- 100 mM $CaCl_2$
- Selective antibiotics[b]

Protocol 1. *Continued*

Method

1. Inoculate 50 ml of 2YT medium with 0.5 ml of an overnight culture of the respective *E. coli* strain.

2. Grow the culture with shaking at 37 °C to an optical density at 600 nm of 0.3–0.4.

3. Collect the cells by centrifugation at 6000 *g* for 10 min at 4 °C. Resuspend the cells in 20 ml ice-cold 100 mM $CaCl_2$.

4. Repeat step **3** resuspending the cells in 10 ml 100 mM $CaCl_2$.

5. Resuspend the cells in 2 ml ice-cold 100 mM $CaCl_2$ and keep them on ice for at least 15 min before use.

6. For transformation of *E. coli* with phasmid DNA add 50–100 ng phasmid DNA to 200 µl of competent cells. Incubate on ice for 15 min. Transfer to 42 °C for 3 min. Add 1 ml 2YT medium and incubate at 37 °C for 60 min. Streak out 200 µl on a 2YT agar plate containing an appropriate selective antibiotic.

[a] BMH71–18/*mutS* strain genotype: Δ(*lac-pro*AB), *supE*, *thi*, *mutS*::Tn 10 (Tet[r]); F′ *lacI*[q], *lacZ*ΔM15, *proA*[+]B[+].

[b] Ampicillin (1000 × stock solution) 100 mg/ml of ampicillin sodium salt in water. Sterilize by filtration (0.2 µm pore size) and store at −20 °C. Chloramphenicol (1000 × stock solution) 25 mg/ml in 100% ethanol. Store at −20 °C. Tetracyclin (1000 × stock solution) 20 mg/ml of tetracyclin in 70% (v/v) ethanol. Store at −20 °C.

Helper phage M13KO7 (21) is by itself a phasmid (also called phagemid) because it contains both the M13 and a p15A plasmid replication origin. M13KO7 also contains a kanamycin resistance marker from Tn903. *Protocol 2* describes the determination of phage titres and the preparation of phage stocks. Phage stock solutions should have titres in the range of 10^{11}–10^{12} plaque-forming units (p.f.u.) per millilitre. Phage stocks may be stored at 4 °C. As a starting material for the preparation of a M13KO7 phage stock, a very small plaque should be used to ensure maintenance of the reduced efficiency of phage propagation (21).

Protocol 2. M13 phage stock preparation

Equipment and reagents

- Host *E. coli* strain, e.g. BMH71–18[a] or NM522[b]
- Phage stock (e.g. M13mp9 or M13KO7)
- 2YT medium (*Protocol 1*)
- 2YT agar plates: 16 g tryptone, 10 g yeast extract, 15 g Bacto agar, 5 g NaCl, made up with water to 1 litre and sterilized by autoclaving

- Glucose minimal medium (*Protocol 1*)
- Top agar: 10 g tryptone, 5 g NaCl, 6.5 g Bacto agar, made up with water to 1 litre and sterilized by autoclaving
- Sorvall SS34 centrifuge tubes or equivalent
- Laboratory film (Parafilm 'M')
- Sterile toothpicks

Method

1. Make serial 100-fold dilutions of a phage stock (e.g. M13mp9 or M13KO7) from 10^2-fold to 10^8-fold in sterile tubes using 2YT medium.

2. Melt top agar in a bath of boiling water and pipette 2.5 ml aliquots into four sterile tubes. Keep at 45 °C until use.

3. Place 0.2 ml of an overnight culture of a male *E. coli* strain (e.g. BMH71–18) grown in glucose minimal medium into each of the four sterile tubes and add to each 0.1 ml of the 10^{-4}, 10^{-6}, and 10^{-8} phage dilution. The fourth tube serves as a control. Mix thoroughly and pour each sample on to a 2YT agar plate pre-warmed at 37 °C. After the top agar has solidified at room temperature, incubate the plates upside down at 37 °C overnight.

4. Transfer a single plaque into 50 ml of 2YT medium using a sterile toothpick. Incubate on a shaking platform overnight at 37 °C.

5. Transfer the culture to a centrifuge tube and pellet the cells by centrifugation for 10 min at 25 000 g in a Sorvall centrifuge using a SS34 rotor or equivalent conditions. Transfer the supernatant to a fresh tube and repeat the centrifugation step. Fill sterile 1.5 ml microcentrifuge tubes with 1 ml aliquots of the supernatant and seal the lids with laboratory film (Parafilm 'M'). Store at 4 °C. Phage from this stock should be viable for at least one year.

6. To determine the phage titre, repeat steps **1–3**. Count the plaques the following morning. The titre is the number of plaques × 10 × dilution factor (p.f.u/ml).

[a] BMH71–18 strain genotype: Δ(*lac-pro*AB), *sup*E, *thi*; F' *lac*I^q, *lac*ZΔM15, *pro*A$^+$B$^+$.
[b] NM522 strain genotype: Δ(*lac-pro*AB), *sup*E, *thi*, Δ(*hsd*MS-*mcrB*); F' *lac*I^q, *lac*ZΔM15, *pro*A$^+$B$^+$.

2.4 Vectors

2.4.1 Phage M13

A prerequisite of oligonucleotide-directed mutation construction via a heteroduplex DNA intermediate is that the target DNA is available in single-stranded form. The life cycle of filamentous phage M13 has been exploited extensively in the production of single-stranded template DNA (22, 23). M13 double-stranded replicative form (RF) DNA can be conveniently isolated from phage-infected *E. coli* cells and manipulated in the same way as plasmid DNA. Phage M13 replicates by a rolling circle mechanism, which eventually results in the continuous synthesis of the circular (+) strand DNA. Circular (+) strand DNA becomes encapsidated into phage particles which are released into the culture medium, from where they can be easily isolated. A

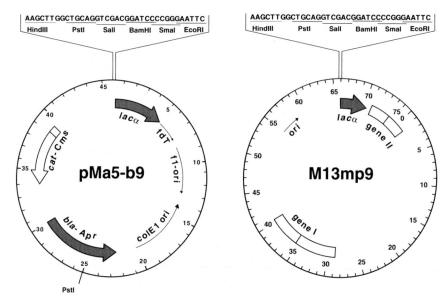

Figure 5. Genetic map of pMa5-b9 and M13mp9. The nucleotide sequence of the poly-linker is shown above the respective map. The position of the β-lactamase coding region (*bla*), the gene coding for chloramphenicol acetyltransferase (*cat*), the *lacZ* gene α fragment (*lacα*), the colE1 origin of replication (colE1 ori), the filamentous phage origin (f1-ori), and the phage fd transcription termination signal (fdT) are indicated. In all cases the *arrow* denotes the functional orientation of the genetic element.

number of derivatives of filamentous phage M13 have been constructed, which facilitate the recognition of clones carrying foreign DNA fragments by the colour (blue/colourless) of the plaques produced. The M13mp (23, 24) series contains various polylinkers within the 5′ terminal portion of the *lacZ* gene α fragment, which is able to complement a chromosomally located *lacZ*ΔM15 mutation. When there is no foreign DNA inserted into this vector, a functional α peptide is produced that together with the LacZΔM15 gene product produces a functional β-galactosidase. Enzymatic activity can be visualized on indicator plates containing the inducer isopropyl-β-D-thiogalactopyranoside (IPTG) and the chromogenic substrate 5-bromo-4-chloro-3-indolyl-β-D-thiogalactopyranoside (Xgal). The cleavage of Xgal by β-galactosidase generates a blue plaque colour. On the other hand, a cloned insert, which interrupts the *lacα* gene leads to a colourless plaque phenotype. A genetic map of M13mp9 is shown in *Figure 5*.

2.4.2 Phasmids

During recent years, the properties of plasmids and phages have been combined through the construction of phasmid (also called phagemid) vectors, which contain both a plasmid and a filamentous phage replication origin (25, 26). These phasmids can be propagated and handled as plasmids in the customary way. However, upon infection with a helper phage such as M13KO7 (21), rolling circle replication is initiated by the gene II product (27) provided by the incoming phage. As a result, one phasmid DNA strand becomes specifically packaged into coat protein synthesized under control of the helper phage and is released together with helper phage into the culture medium as phage-like particles.

A number of special purpose phasmid vectors have been developed. These include:

(a) The pMa/c series for directed mutagenesis using the gapped duplex approach (11).

(b) The pBluescript vector series (Stratagene Cloning Systems; ref. 28).

(c) The pT7T3 series (Pharmacia LKB Biotechnology) for *in vitro/in vivo* transcription of the target gene under control of a T7/T3 promoter.

(d) The pUC118/pUC119 phasmids (21) for blue/colourless colony pheno-type assay via α-complementation.

Practically any phage or phasmid vector may be used for the phosphoro-thioate method and the deoxyuridine method. The series of pMa/c phasmid vectors has been constructed for directed mutagenesis using the gapped duplex method. These contain appropriate expression signals and also allow strand selection using alternating selectable markers (ref. 11; *Figure 2*).

2.4.3 Preparation of single-stranded vector DNA

The methods described in *Protocols 3* and *4* allow the respective isolation of phage or phasmid DNA that is suitable for use in the various mutagenesis procedures. If the single-stranded DNA is to be used for the deoxyuridine mutagenesis scheme, it is important that the template DNA is free of small contaminating fragments of DNA or RNA, which may act as primers in the polymerization reaction. This could lead to the formation of homoduplex DNA without incorporation of the mutagenic oligonucleotide. Single-stranded phasmid DNA seems to be more prone to be contaminated by such fragments than phage DNA. To generate sufficient amounts of phasmid particles, it is often necessary to cultivate the infected cells over-night until they have reached the stationary growth phase. Under these conditions cell lysis occurs to a significant degree, and this may lead to the accumulation of DNA or RNA derived from lysed cells in the culture medium. This DNA/RNA may co-purify in the course of the isolation of phage/phasmid particles, and can therefore contaminate the single-stranded DNA preparation.

Protocol 3. Preparation of virion DNA of recombinant M13mp9

Equipment and reagents

- Overnight culture of *E. coli* BMH71–18[a] in glucose minimal medium (*Protocol 1*)
- M13mp9 phage stock
- 2YT medium (*Protocol 1*)
- TES buffer: 100 mM Tris–HCl pH 8.0, 300 mM NaCl, 1 mM EDTA
- TE buffer: 10 mM Tris–HCl pH 8.0, 0.1 mM EDTA
- 20% (w/v) PEG 6000, 2.5 M NaCl

- DNase-free RNase A
- Phenol, TE-saturated
- Phenol/chloroform (1 : 1)
- Chloroform/isoamyl alcohol (24 : 1)
- 3 M sodium acetate
- Ethanol
- Sorvall SS34 centrifuge tubes or equivalent
- 20 ml glass pipettes
- Quartz cuvette

Method

1. Inoculate 50 ml of 2YT medium with 250 μl of a fresh overnight culture of *E. coli* strain BMH71–18 grown in glucose minimal medium. Infect the culture with 50 μl of the phage stock (*c.* 10^{12} p.f.u/ml; see *Protocol 2*) of the recombinant M13mp9 clone. Incubate on a shaking platform overnight at 37°C.

2. Transfer 2 × 25 ml of the culture to two centrifuge tubes and pellet the cells by centrifugation for 10 min at 25 000 g in a Sorvall centrifuge using a SS34 rotor or equivalent conditions.

3. Carefully remove 16 ml of the supernatant with a 20 ml glass pipette and transfer it to a fresh tube. Take care not to transfer any cells. Also avoid co-transferring the white smear of lipids derived from the lysed cells from the top of the liquid phase.

4. Add 1/4 vol. of 20% (w/v) PEG 6000 in 2.5 M NaCl. Allow the phage particles to precipitate for 30 min at 4°C.

5. Centrifuge at 25 000 g for 10 min at 4°C in a Sorvall centrifuge using a SS34 rotor or equivalent conditions. Discard the supernatant and remove remaining traces with a drawn-out Pasteur pipette.

6. Add 300 μl TES buffer and 3 μl of DNase-free RNase A (10 mg/ml). Incubate for 30 min at 37°C.

7. Centrifuge for 5 min in a microcentrifuge and transfer the supernatant to another tube. Take care not to co-transfer the sedimented debris.

8. Add 300 μl of TE-saturated phenol, and mix by vortexing for 15–20 sec. Spin for 2 min in a microcentrifuge. Transfer the aqueous (upper) phase to a fresh tube. Repeat the extraction with phenol/chloroform (1 : 1), then with chloroform/isoamyl alcohol (24 : 1).

9. Precipitate the liberated single-stranded DNA by adding 1/10 vol. 3 M sodium acetate and 3 vol. of ethanol. Store at −20°C for 45 min.

10. Centrifuge in a microcentrifuge for 5 min. Discard the supernatant and remove remaining traces with a drawn-out Pasteur pipette. Dissolve the pellet in 100 μl TE buffer.

11. Take a 10 μl sample, dilute to 1 ml, and determine the optical density at 260 nm using a 1 ml quartz cuvette. 1 OD_{260} corresponds to approximately 36 μg/ml of single-stranded DNA (ss DNA). Typically about 50–100 μg ss DNA is obtained.

[a] BMH71–18 strain genotype:Δ (*lac-proAB*), *supE, thi;* F' *lacl*[q], *lacZ*ΔM15, *proA*[+]B[+].

Protocol 4. Preparation of single-stranded phasmid DNA

Reagents

- Overnight culture in glucose minimal medium (*Protocol 1*) of *E. coli* BMH71–18,[a] NM522,[b] or BW313[c]
- Stock of helper phage M13K07 (*Protocol 2*)
- 2YT medium (*Protocol 1*)
- 100 mM $CaCl_2$
- Selective antibiotics[d]

Method

1. Prepare an overnight culture of an appropriate *E. coli* strain (e.g. BMH71–18, NM522, or BW313) in glucose minimal medium. Starting from the overnight culture, prepare competent cells and transform with the phasmid (*Protocol 1*).

2. The next day, pick a single colony and transfer to 3 ml 2YT containing the appropriate antibiotic for phasmid selection. Infect the culture with 5 μl helper phage M13K07 (c. 10^{12} p.f.u./ml; see *Protocol 2*) and incubate at 37 °C for 2 h.

3. Place 50 ml 2YT selective media containing kanamycin plus the appropriate antibiotic in a 100 ml Erlenmeyer flask. Inoculate with 2 ml of the infected culture. Incubate overnight on a shaking platform with good aeration at 37 °C.

4. To isolate ss DNA follow *Protocol 3*, steps 2 to 11.

[a] BMH71–18 strain genotype: Δ(*lac-proAB*), *supE, thi;* F' *lacl*[q], *lacZ*ΔM15, *proA*[+]B[+].
[b] NM522 strain genotype: Δ(*lac-proAB*), *supE, thi,* Δ(*hsd*MS-*mcr*B); F' *lacl*[q], *lacZ*ΔM15, *proA*[+]B[+].
[c] BW313 strain genotype: HfrKL16 PO/45 (*lys*A61–62), *dut*1, *ung*1, *thi*1, *rel*A1.
[d] Ampicillin (1000 × stock solution) 100 mg/ml of ampicillin sodium salt in water. Sterilize by filtration (0.2 μm pore size) and store at − 20 °C. Chloramphenicol (1000 × stock solution) 25 mg/ml in 100% ethanol. Store at − 20 °C. Kanamycin (1000 × stock solution) 75 mg/ml in water. Sterilize by filtration (0.2 μm pore size) and store at − 20 °C.

2.5 Synthetic oligonucleotides

2.5.1 General considerations

The mutagenic primer consists of two flanking sequences complementary to the target DNA, which must provide a site-specific and stable hybridization, and a core sequence to be left unpaired after annealing. For a sequence of normal G/C content and an unpaired core of just one nucleotide these flanking regions should each be about eight or nine nucleotides long. For substitution or insertion of longer DNA stretches or deletion of large regions in the template the flanking sequences should be in the range of 12–15 nucleotides. To simplify the screening procedure for detecting the clone carrying the desired mutation, it is advisable to include a restriction enzyme cleavage site within the primer core sequence whenever possible. This allows fast identification of clones carrying the desired mutation by restriction analysis. Alternatively, if a unique restriction site is removed during the mutagenesis reaction, the wild-type DNA can be selectively eliminated from the mixed pool of wild-type and mutant DNA by restriction endonuclease cleavage and re-transformation (29). In many cases, the degeneracy of the genetic code allows the introduction of such a cleavage site in reasonable proximity to the position of the desired mutation without any change in the coded amino acid sequence. A number of computer programs (e.g. the 'MAP/SILENT' option in the UWGCG program package; ref. 30) have been devised to identify appropriate 'silent' cleavage sites.

2.5.2 Purification and phosphorylation

Chemical homogeneity of the mutagenic oligonucleotide is a prerequisite for the construction of specific mutations. Two purification schemes have proven successful:

- high pressure liquid chromatography (HPLC) on reverse phase (31)
- preparative acrylamide gel electrophoresis (32).

Current methods for automated oligonucleotide synthesis yield oligonucleotides with a 5' terminal dimethoxytrityl protective group. This group provides a convenient handle for HPLC on a reverse C18 stationary phase (*Protocol 5*).

Protocol 5. HPLC purification of oligonucleotides

Equipment and reagents

- HPLC loading buffer: 100 mM triethyl-ammonium acetate pH 7.0, 10% (v/v) acetonitrile
- Mobile phases A and B: mobile phase A is 100 mM triethylammonium acetate, adjusted to pH 7.0 with acetic acid; mobile phase B is acetonitrile
- TE buffer: 10 mM Tris–HCl pH 8.0, 0.1 mM EDTA
- 80% (v/v) acetic acid
- Ethyl acetate
- 3 M sodium acetate
- Ethanol
- Quartz cuvette

Method

1. The oligonucleotide (0.1 μmol), which has been synthesized 'Trityl-on' on an automated DNA synthesizer is normally obtained in 25% (v/v) ammonia. Distribute the solution into four 1.5 ml microcentrifuge tubes and freeze at −80 °C for 30 min. Lyophilize the samples (e.g. in a Speed-Vac concentrator).

2. Dissolve each of the lyophilized oligonucleotide aliquots in 100 μl of HPLC loading buffer. Mix by vortexing and centrifuge for 5 min in a microcentrifuge. Carefully remove the cleared solution from the sedimented material and transfer it to a new microcentrifuge tube.

3. The HPLC system should be equipped with a UV detector operating at 260 nm. Dual wavelength monitoring (260/280 nm) or diode array detectors are convenient but not necessary. Run the column (C18 reverse phase silica gel; bed dimensions: 250 mm × 4 mm; particle size: 5 μm) with the following gradient of mobile phases A and B at a flow rate of 2 ml/min:

min	% mobile phase B
0–1.5	10
1.5–2	10–15.5
2–23	15.5–34
23–25	34–40
25–27	40–10
27–30	10

Column ready for next injection.

4. Inject 10 μl for analytical and 100 μl for preparative scale purification. The oligonucleotide normally elutes after approximately 12 min (26% mobile phase B). Collect the fractions from the first sharp increase in the optical density up to the peak maximum. Discard the fractions corresponding to the tail of the peak, as these are often contaminated with shorter oligonucleotides.

5. Freeze the pooled fractions at −80 °C for 30 min and lyophilize. For removal of the 5′ dimethoxytrityl group, dissolve lyophilized fractions in 100 μl 80% (v/v) acetic acid and keep at room temperature for 15 min. Freeze at −80 °C for 30 min and lyophilize.

6. Dissolve the pellet in 100 μl TE buffer. Add 100 μl ethyl acetate, mix by vortexing, and centrifuge for 2 min at room temperature in a microcentrifuge. Remove the organic (upper) phase. Repeat the extraction twice. Transfer the aqueous phase into a new microcentrifuge tube and precipitate the DNA by adding 1/10 vol. 3 M sodium acetate and 3 vol. ethanol. Store at −20 °C for 60 min.

7. Centrifuge at 25 000 *g* for 5 min at 4 °C. Discard the supernatant and

Protocol 5. *Continued*

dissolve the pellet in 100 μl TE buffer. Take a 5 μl sample, dilute to 1 ml, and read the optical density at 260 nm in a 1 ml quartz cuvette of 1 cm path length. 1 OD_{260} corresponds to approximately $10^5/n$ pmol of the synthetic oligonucleotide, where n is the number of monomers that make up the mutagenic primer.

After enzymatic phosphorylation, the purified oligonucleotide can be directly used for the mutagenesis experiment. Alternatively, the oligonucleotide is detritylated either manually (*Protocol 5*) or automatically, and then enzymatically phosphorylated (*Protocol 6*) prior to purification via preparative polyacrylamide gel electrophoresis (*Protocol 7*).

If the oligonucleotide contains the 5′-dimethoxytrityl protective group (DNA synthesizer option 'Trityl-on') lyophilize it and follow *Protocol 5*, steps 5 to 7. In case of automatic removal of the 5′ dimethoxytrityl group (DNA synthesizer option 'Trityl-off'), the faster procedure given in *Protocol 6* may be used for the removal of the alkali-labile protective groups which are cleaved off during ammonia treatment of the oligonucleotide.

Protocol 6. Phosphorylation of the mutagenic primer

Equipment and reagents

- 10 × kinase buffer: 500 mM Tris–HCl pH 7.5, 100 mM $MgCl_2$, 10 mM dithiothreitol
- TE buffer: 10 mM Tris–HCl pH 8.0, 0.1 mM EDTA
- 100 mM ATP
- n-Butanol
- 1 M ammonium acetate
- Ethanol

- Polynucleotide kinase (Boehringer-Mannheim Biochemicals)
- Phenol, TE-saturated
- Phenol/chloroform (1 : 1)
- Chloroform/isoamyl alcohol (24 : 1)
- Sorvall SS34 rotor tubes or equivalent
- Quartz cuvette

Method

1. Add 10 vol. *n*-butanol to the solution of the detritylated oligonucleotide in 25% (v/v) ammonia. Vortex briefly, and centrifuge at 25 000 *g* for 5 min in a Sorvall SS34 rotor (or equivalent) at room temperature.

2. Decant the supernatant and dissolve the pellet in 1 ml TE buffer. Add 10 ml *n*-butanol and repeat the centrifugation step.

3. Dissolve the pellet in 200 μl 1 M ammonium acetate. Add 600 μl ethanol, mix briefly by vortexing, and centrifuge in a microcentrifuge for 5 min. Discard the supernatant and resuspend the pellet in 1 ml 70% (v/v) ethanol. Re-centrifuge and carefully remove the supernatant. Dissolve the pellet in 100 μl TE buffer and determine the DNA concentration as described in *Protocol 5*, step 7.

4. Prepare the following mixture:

- oligonucleotide 2–5 nmol
- 100 mM ATP 2 μl
- 10 × kinase buffer 10 μl
- H_2O to 100 μl final volume
- polynucleotide kinase 10 U

Incubate for 60 min at 37 °C.

The mixture can be used directly for purification by polyacrylamide gel electrophoresis (*Protocol 7*). For HPLC purified oligonucleotides a gel electrophoresis purification step can be omitted. In this case follow step 5.

5. Add to the phosphorylation mixture 300 μl of TE-saturated phenol. Mix by vortexing for 15–20 sec and centrifuge for 2 min in a microcentrifuge. Transfer the aqueous (upper) phase to a fresh tube. Repeat the extraction with phenol/chloroform (1:1), then with chloroform/isoamyl alcohol (24 : 1).

6. Precipitate the DNA by adding 1/10 vol. 3 M sodium acetate and 3 vol. ethanol. Store at −20 °C for 45 min.

7. Centrifuge for 5 min in a microcentrifuge. Discard the supernatant and remove remaining traces with a drawn-out Pasteur pipette. Dissolve the pellet in 100 μl TE buffer. Determine the DNA concentration as described in *Protocol 5*, step 7.

Protocol 7. Purification of oligonucleotides by polyacrylamide gel electrophoresis

Equipment and reagents

- Polyacrylamide gel electrophoresis equipment
- 10 × TBE buffer: 108 g Tris base, 55 g boric acid, 40 ml 50 mM EDTA pH 8.0, H_2O to a final volume of 1 litre
- Formamide dye mix: 0.1 g bromophenol blue, 0.1 g xylene cyanol FF, 2 ml 0.5 M EDTA pH 8.0, 100 ml formamide
- Formamide
- Urea
- Acrylamide stock solution: 30% acrylamide (w/v)/0.8% bisacrylamide (w/v)
- 25% (w/v) ammonium peroxodisulfate

- TEMED (*N,N,N′N′*-tetramethylethylenediamine)
- Phenol, TE-saturated
- Phenol/chloroform (1 : 1)
- Chloroform/isoamyl alcohol (24 : 1)
- 3 M sodium acetate
- Ethanol
- Fluorescent thin-layer chromatographic plate (e.g. Merck 60F$_{254}$)
- Saran-Wrap
- Hand-held ultraviolet lamp (254 nm)

Method

1. Prepare the following solution:

- urea 48 g
- 10 × TBE 10 ml

H. Kolmar and H.-J. Fritz

Protocol 7. *Continued*

- 30% acrylamide (w/v)/0.8% bis-   33–50 ml
  acrylamide (w/v) stock solution
- H_2O to 100 ml final volume.

For oligonucleotides shorter than 30 nucleotides use a 15% acrylamide gel (50 ml acrylamide stock solution). For oligonucleotides larger than 30 nucleotides use a 10% gel (33 ml acrylamide stock solution). Start the polymerization by the addition of 100 μl 25% (w/v) ammonium peroxodisulfate and 100 μl TEMED, and cast a gel with dimensions 20 × 20 × 0.2 cm. Combs can be inserted to form up to 15 gel slots.

2. Carefully rinse the gel slots with TBE using a 10 ml plastic syringe. Add 1/3 vol. formamide to each oligonucleotide sample and apply it to the gel with a needle-tipped syringe. Do not apply more than 50 μl to a gel pocket of 1 cm width. Apply 20 μl formamide dye on to a separate slot. Run the gel at 20 W until the bromophenol blue dye (dark blue) has migrated approximately two thirds of the gel length. Remove glass plates.

3. Cover a fluorescent thin-layer chromatographic plate (e.g. Merck 60F$_{254}$) with Saran-Wrap. Place the gel on top and illuminate with a hand-held ultraviolet lamp (254 nm).

4. Cut out the slowest migrating band. Transfer the gel slice to a micro-centrifuge tube. Add 500 μl TE buffer and incubate overnight at 65 °C.

5. Centrifuge for 5 min in a microcentrifuge. Transfer the supernatant to a new microcentrifuge tube. Add 1 vol. TE-saturated phenol, and mix by vortexing for 15–20 sec. Centrifuge for 2 min in a microcentrifuge. Transfer the aqueous (upper) phase to a fresh tube. Repeat the extraction with phenol/chloroform (1 : 1), then with chloroform/isoamyl alcohol (24 : 1).

6. Precipitate the DNA by adding 1/10 vol. 3 M sodium acetate and 3 vol. ethanol. Store at −20 °C for 45 min. Centrifuge in a microcentrifuge at 4 °C. Discard the supernatant and resuspend the pellet in 100 μl TE buffer. Determine the DNA concentration as described in *Protocol 5*, step 7.

2.5.3 Oligonucleotides containing phosphorothioate linkages

Oligonucleotides with one or two internucleotide phosphodiester bridges at their 5′ terminus replaced by a phosphorothioate linkage are useful reagents in the mix-heat-transform (MHT) procedure (refs 10, 20; see also Sections 1.5 and 5.1). The phosphorothioate bridge markedly increases the yield of synthetic marker, apparently by providing protection against 5′ to 3′ exonucleolytic attack on the oligonucleotide *in vivo*. The phosphorothioate bridge can be introduced during the automated phosphoramidite-type of oligonucleotide synthesis with elemental sulfur dissolved in carbon disulfide/acetonitrile (20) or, even more conveniently, with tetraethylthiuram disulfide (Applied Biosystems Inc.) as the sulfurizing agent.

3. The gapped duplex DNA procedure

The basic principle of the gapped duplex method was outlined in Section 1.1 and in *Figure 2*. It was originally developed for the oligonucleotide-directed mutagenesis of a target fragment cloned in M13mp9 phage (9, 10). More recently, it has been adapted for the mutagenesis of DNA carried in phasmid vectors of the pMa/c series (11). The basic methodologies are quite similar and typical marker yields in the range of 50–90% are obtained with both procedures.

3.1 Use of phage M13mp9/M13mp9rev as cloning vector

The genetic selection in the gapped duplex DNA approach using the two different filamentous phages M13mp9 and M13mp9rev is based on the presence of two amber codons, which have been introduced into genes I and II of M13mp9. As a result, M13mp9 can only propagate in an *Escherichia coli* amber suppressor strain. The target DNA is cloned into this type of phage and single-stranded DNA of the recombinant phage is isolated as described in *Protocol 3*. Gapped duplex DNA is formed by strand transfer from linearized double-stranded DNA of phage M13mp9rev. M13mp9rev is a derivative of M13mp9, in which the two amber codons are replaced by sense codons. Thus, propagation of this phage does not require an amber suppressor host. This allows selection for the progeny of the shorter strand of the gd DNA strand. Note that it is this strand to which the mutagenic oligonucleotide becomes physically linked.

The size of the gap has a significant, yet not overwhelming influence in marker yield (9). Therefore, the large *Hin*dIII/*Eco*RI fragment of M13mp9rev with the complete polylinker region removed—see below and *Protocol 8*—can be used for gapped duplex DNA formation with any type of recombinant single-stranded DNA of M13mp9, irrespective of the restriction site(s) that have been used for cloning of the target fragment.

For the preparation of M13mp9rev double-stranded DNA (ds DNA), any medium scale procedure can be applied. The portion of nicked DNA in the preparation of ds DNA should be as small as possible. Hybridization of fragments resulting from a nick in the complementary strand of the double-stranded DNA might result in heteroduplex DNA molecules with an undesirably large gap.

It is advisable, yet not absolutely necessary, to remove the small polylinker fragment resulting from the *Hind*III/*Eco*RI cleavage of M13mp9rev by sucrose gradient centrifugation (*Protocol 8*). This purification procedure is also advantageous, because it allows the removal of remaining traces of uncleaved vector DNA, which can lead to unwanted background.

Protocol 8. Preparation of the large *Hind*III/*Eco*RI restriction fragment of M13mp9rev

Equipment and reagents

- Overnight culture of *E. coli* strain BMH71–18[a] in glucose minimal medium (*Protocol 1*)
- M13mp9rev phage stock
- 2YT medium (*Protocol 1*)
- Appropriate restriction enzymes (e.g. *Hind*III and *Eco*RI)
- Sucrose gradient buffer: 100 mM Tris–HCl pH 8.0, 10 mM EDTA, 100 mM NaCl, 20 μg/ml ethidium bromide
- TE buffer: 10 mM Tris–HCl pH 8.0, 0.1 mM EDTA

- Cleavage buffer: 50 mM Tris–HCl pH 7.5, 10 mM $MgCl_2$, 1 mM dithiothreitol
- Phenol, TE-saturated
- Phenol/chloroform (1 : 1)
- Chloroform/isoamyl alcohol (24 : 1)
- 3 M sodium acetate
- Ethanol
- Needle-tipped syringe (1 ml)
- Ultra Clear centrifuge tubes (Beckman SW41, or equivalent)
- Type SW41 rotor (Beckman or equivalent)

Method

1. Inoculate two flasks containing 100 ml 2YT medium each with 5 ml of a fresh overnight culture of strain BMH71–18 grown in glucose minimal medium. Infect each culture with 0.5 ml of M13mp9rev stock solution (for the preparation of M13 stock solutions see *Protocol 6*). Shake the culture flask at 37°C for 6 h.

2. Collect the cells by centrifugation and prepare RF DNA by following any protocol for the medium scale preparation of plasmid DNA (e.g. ref. 33).

3. Incubate 25 μg double-stranded M13mp9rev DNA with 50 U of the appropriate restriction endonucleases (e.g. *Hind*III and *Eco*RI) in 300 μl cleavage buffer at 37°C for 60 min. Place the reaction mixture at 65°C for 10 min. Withdraw 10 μl for electrophoretic analysis.

4. Prepare a solution of 18.5% (w/v) sucrose in sucrose gradient buffer. Transfer the solution to Ultra Clear centrifuge tubes (Beckman SW41, or equivalent) and freeze at −20°C.

5. Thawing at room temperature leads to formation of a sucrose gradient. Overlay this gradient with the DNA solution resulting from the restriction digest.

6. Centrifuge at 30 000 r.p.m for 14–16 h in a type SW41 rotor (Beckman or equivalent) at 15°C.

7. Harvest the band consisting of the large restriction fragment (which should be visible under UV light in the central part of the tube) with a

needle-tipped syringe, pierced through the wall of the tube. Collect the DNA in a volume of about one ml.

8. Add 100 μl 3 M sodium acetate and 3 ml cold ethanol and store at −20 °C for 30 min. Centrifuge at 4 °C for 10 min in a microcentrifuge.

9. Re-dissolve the pellet in 300 μl TE buffer, add 300 μl of TE-saturated phenol, and mix by vortexing for 15–20 sec. Centrifuge for 2 min in a microcentrifuge. Transfer the aqueous (upper) phase to a fresh tube. Repeat the extraction with phenol/chloroform (1 : 1) and then with chloroform/isoamyl alcohol (24 : 1).

10. Precipitate the DNA by adding 1/10 vol. 3 M sodium acetate and 3 vol. ethanol. Store at −20 °C for 45 min. Centrifuge for 5 min in a micro-centrifuge. Discard the supernatant and remove remaining traces with a drawn-out Pasteur pipette. Dissolve the pellet in 100 μl TE buffer.

[a] BMH71–18 strain genotype: Δ(*lac-pro*AB), *sup*E, *thi*; F' *lac*I[q], *lacZ*ΔM15, *pro*A[+]B[+].

The mutagenic oligonucleotide is annealed to the single-stranded target region of the gd DNA. The remaining gaps are filled and sealed *in vitro* using DNA polymerase and DNA ligase. This results in the integration of the muta-genic primer into the DNA strand derived from M13mp9rev (*Protocol 9*). The commercially available DNA polymerase that has the most favourable properties for complementary strand synthesis in site-directed mutagenesis is native T7 DNA polymerase, the T7 gene 5 product complexed with *E. coli* thoredoxin (15).

The strain BMH71–18/*mutS* (DNA mismatch repair deficient) is used as the transfection host for the heteroduplex DNA. This ensures genetic linkage of the oligonucleotide-mediated mutation to the selectable M13mp9rev marker. Marker segregation and selection for the progeny of M13mp9rev is achieved by re-infection at very low multiplicity of a non-suppressor strain with the mixed phage progeny of the primary transfection.

Protocol 9. Gapped duplex DNA mutagenesis using M13mp9/ M13mp9rev

Reagents

- *E. coli* BMH71–18/*mutS*[a] competent cells (*Protocol 1*)
- Overnight culture of *E. coli* su⁻ strain (e.g. WK6[b]) in glucose minimal medium (*Protocol 1*)
- Recombinant M13mp9 ss DNA (*Protocol 3*)
- M13mp9rev DNA fragment (*Protocol 8*)
- 2YT medium (*Protocol 1*)
- 2YT agar plates (*Protocol 1*)

- Top agar (*Protocol 1*)
- 5' Phosphorylated oligonucleotide (*Protocols 5–7*)
- T7 DNA polymerase form II unmodified (New England Biolabs)
- T4 DNA ligase (Boehringer-Mannheim Biochemicals)
- 10 × annealing buffer: 200 mM Tris–HCl pH 7.4, 20 mM MgCl₂, 500 mM NaCl

Protocol 9. *Continued*

- Dilution buffer: 20 mM potassium phosphate pH 7.4, 1 mM dithiothreitol, 0.1 mM EDTA, 55% (v/v) glycerol
- TAE buffer: 40 mM Tris–acetate, 2 mM EDTA, pH 8.0

- 10 × synthesis buffer: 200 mM Tris–HCl pH 7.4, 100 mM MgCl$_2$, 20 mM dithiothreitol, 4 mM ATP, 5 mM of each dNTP (N = G,A,T,C)
- 1% (w/v) agarose in TAE buffer containing 5 µg/ml ethidium bromide

Method

1. Mix 0.1 pmol (approximately 300 ng) of the purified linearized fragment derived from M13mp9rev with 0.5 pmol (approximately 1 µg) ss DNA of M13mp9 containing the target fragment. Add 4 µl 10 × annealing buffer and adjust the volume to 40 µl with H$_2$O.

2. Incubate the reaction mixture at 100 °C for 2 min. Place at 65 °C for 5–10 min.

3. Analyse a 15 µl aliquot together with 200 ng of the ss DNA and 100 ng of the vector fragment by electrophoresis through a 1% agarose gel in 1 × TAE containing 5 µg/ml ethidium bromide. Gapped duplex DNA has approximately (depending on the size of the gap) the same mobility as form II (i.e. double-stranded open circle) phage DNA.

4. If necessary, the procedure can be interrupted at this point by storing the rest of the hybridization mixture at −20 °C.

5. Mix 8 µl of the hybridization mixture and 4–8 pmol (2 µl) of the phosphorylated oligonucleotide in a microcentrifuge tube. Incubate at 70 °C for 5 min. Allow the reaction mixture to cool in the water-bath over a period of approximately 30 min to room temperature. Transfer the reaction mixture to an ice bath.

6. Dilute unmodified T7 DNA polymerase (the T7 gene 5 protein/thioredoxin complex) in dilution buffer to a final concentration of 1 U/µl.

7. To the reaction mixture add:
   - 10 × synthesis buffer   3 µl
   - H$_2$O   14 µl
   - T4 DNA ligase (1 U/µl)   2 µl
   - diluted T7 DNA polymerase (1 U/µl)   1 µl.

8. Incubate the mixture sequentially on ice for 5 min, at 25 °C for 5 min, and finally at 37 °C for 60 min. Transfer the mixture back to an ice bath.

9. Prepare competent cells from an overnight culture of a male *E. coli* strain deficient in DNA mismatch repair (e.g. BMH71–18/*mutS*) as described in *Protocol 1*. Add the reaction mixture to 200 µl of competent cells. Incubate on ice for at least 15 min.

10. Melt top agar in a bath of boiling water and pipette 2.5 ml portions into a sterile tube. Keep at 45 °C until use. Add 0.1 ml of the BMH71–

18/*mutS* overnight culture. Incubate the transfection mixture at 42 °C for 3 min, and then add 1 ml 2YT medium. Transfer 200 µl to the tube of top agar, mix thoroughly, and pour on to a 2YT agar plate pre-warmed at 37 °C. Incubate the plate upside down at 37 °C overnight. Determine titre of plaque-forming units in the transfection mixture.

11. The remainder of the suspension (approximately 1 ml) is used to inoculate 50 ml of 2YT medium. Incubate with agitation overnight at 37 °C.

12. The mixed phage population obtained is allowed to segregate by re-infecting a su⁻ (non-suppressor) strain such as WK6. To this end, make serial 100-fold dilutions of phage overnight culture from 10^2-fold to 10^8-fold in sterile tubes using 2YT medium. Dilute the 10^8-fold diluted sample an additional ten-fold. Melt top agar in a bath of boiling water and pipette 2.5 ml portions in three sterile tubes. Keep at 45 °C until use.

13. Place 0.2 ml of an overnight culture of an *E. coli* non-suppressor strain (WK6) grown in minimal medium into each of the three sterile tubes and add to each 0.1 ml of the 10^{-6}, 10^{-8}, and 10^{-9} dilution. Mix thoroughly and pour on to a 2YT agar plate pre-warmed at 37 °C. Repeat with other tubes. Incubate the plates upside down at 37 °C overnight.

14. The following morning, inoculate several flasks containing 50 ml 2YT medium each with a single plaque. Incubate on a shaking platform overnight at 37 °C and prepare ss DNA or ds DNA for analysis.

[a] BMH71–18/*mutS* strain genotype: Δ(*lac-pro*AB), *supE*, *thi*, *mutS*::Tn 10 (Tet[r]); F' *lacl*[q], *lacZ*ΔM15, *proA*⁺B⁺.
[b] WK6 strain genotype: Δ(*lac-pro*AB), *thi*, *rpsL* (Str[r]), *nal*[r]; F' *lacl*[q], *lacZ*ΔM15, *proA*⁺B⁺.

3.2 Use of the pMa/c phasmid family as cloning vectors

The pMa/c5-b9 version (34) of the pMa/c phasmid vectors (11) constructed for this method of mutagenesis contains both a colE1 and a f1 phage replication origin, two resistance genes, namely the β-lactamase (*bla*) and chloramphenicol acetyltransferase (*cat*) genes conferring the resistance to ampicillin and chloramphenicol, respectively, and an α complementing region of the *lacZ* gene including a polylinker (*Figure 5*). The vectors contain single amber mutations introduced into either the *bla* or the *cat* gene. Thus, pMa5-b9 confers ampicillin resistance and chloramphenicol sensitivity in a non-suppressor host, while pMc5-b9 confers ampicillin sensitivity and chloramphenicol resistance. For a mutagenesis experiment, the target DNA is cloned into either the pMa or pMc type vector. M13K07 is used as the helper phage in order to generate phage-like particles, from which single-stranded DNA can be isolated (*Protocol 4*). The phasmid vector carrying the alternative functional resistance marker is used for preparation of the linearized (double-stranded) vector fragment. As a standard fragment, the large *Hind*III/*Eco*RI fragment can be isolated according completely with the M13mp9/M13mp9rev system described in *Protocol 8*.

Hybridization of the vector fragment to the single-stranded target DNA results in a gapped duplex DNA, in which different selectable markers are encoded by the two DNA strands. The synthetic oligonucleotide is annealed to this structure and becomes physically linked to the gapped strand via the DNA polymerase/DNA ligase reaction performed *in vitro*. Thus, the desired mutation is coupled to the genetic marker conferred by the gapped strand. The resulting heteroduplex DNA is used for transformation of a mismatch repair deficient (*mutS*) host. The mixed phasmid population is allowed to segregate via transformation of a su⁻ host strain and transformants are selected that confer either chloramphenicol or ampicillin resistance, respectively (*Protocol 10A*). Alternatively, segregation can be performed by phasduction (*Protocol 10B*; ref. 35). In the latter procedure, phage-like particles are produced from the mixed population of primary transformants by helper phage infection. These particles can be used for infection of a non-suppressor recipient strain at a low multiplicity of infection and phasmid-mediated transduction of the synthetic marker. Phasduction can be initiated immediately after transformation with heteroduplex DNA. Therefore, this procedure saves one day compared to the re-transformation procedure. However, due to the high multiplicity of infection of helper phage, clones resulting from the phasduction experiment are in most cases co-infected by the helper phage. This is a welcome event in such cases, where the introduction of the desired mutation is directly screened via nucleotide sequence analysis using ss DNA isolated from single clones.

A single round of mutagenesis results both in the introduction of the desired mutation and in the conversion of the phasmid type from pMa to pMc or vice versa. Thus, a major advantage compared to the gapped duplex method using M13mp9/M13mp9rev is the cycling between two different configurations by the reciprocate selection for ampicillin and chloramphenicol resistance. This allows the introduction of multiple mutations in the course of several consecutive rounds of mutagenesis.

Protocol 10. Gapped duplex DNA mutagenesis using phasmid vectors of the pMa/c series

Equipment and reagents

- *E. coli* BMH71–18/*mutS*[a] competent cells (*Protocol 1*)
- Overnight culture of *E. coli* su⁻ strain (e.g. WK6[b]) in glucose minimal medium (*Protocol 1*)
- Recombinant pMa/c ss DNA (for preparation see *Protocol 4*)
- M13K07 phage stock
- pMa/c vector fragment
- 2YT medium (*Protocol 1*)

- 2YT agar plates (*Protocol 1*) containing chloramphenicol[c] (25 μg/ml) or ampicillin[d] (100 μg/ml), as appropriate
- 2YT agar plates containing streptomycin (50 μg/ml) and chloramphenicol (25 μg/ml) or ampicillin (100 μg/ml), as appropriate (only required for *Protocol 10B*)
- 5′ Phosphorylated oligonucleotide (*Protocols 5–7*)
- T7 DNA polymerase form II unmodified (New England Biolabs)
- T4 DNA ligase (Boehringer-Mannheim Biochemicals)
- 10 × annealing buffer (*Protocol 9*)
- Dilution buffer (Protocol 9)
- 10 × synthesis buffer (*Protocol 9*)
- TAE buffer: 40 mM Tris–acetate, 2 mM EDTA, pH 8.0
- 1% (w/v) agarose in TAE buffer containing 5 μg/ml ethidium bromide

Method

1. Mix 0.1 pmol (approximately 300 ng) of the purified vector fragment with 0.5 pmol (approximately 1 μg) ss DNA of the pMa/c derivative. Add 4 μl 10 × annealing buffer and adjust the volume to 40 μl with H_2O.

2. Follow *Protocol 9*, steps 2 to 8.

3. Prepare competent cells of a DNA mismatch repair deficient *E. coli* strain (e.g. BMH71–18/*mutS* as described in *Protocol 1*). Add the reaction mixture to 200 μl of competent cells. Incubate on ice for 15 min. Transfer to 42 °C for 3 min. Add 1 ml 2YT and incubate at 37 °C for 60 min. Streak out 200 μl on a 2YT agar plate containing chloramphenicol (25 μg/ml) or ampicillin (100 μg/ml), as appropriate.

A. Segregation via re-transformation

1. The remainder of the suspension (approximately 1 ml) is used to inoculate 10 ml of 2YT medium supplemented with the appropriate antibiotic. Incubate on a shaking platform overnight at 37 °C.

2. Use the liquid culture to prepare phasmid DNA following any of the small scale isolation procedures (e.g. ref. 33).

3. Markers of the mixed phasmid DNA population obtained are allowed to segregate by transforming a su⁻ (non-suppressor) strain such as WK6. To minimize the number of double transformants, use a small amount (< 20 ng) of the phasmid DNA preparation. Spread 100 μl aliquots of the transformation mixture on 2YT plates containing the appropriate antibiotic, and incubate overnight at 37 °C.

B. Segregation via phasduction

1. Infect the cell suspension with 5 μl M13KO7 (c. 10^{12} p.f.u./ml). Incubate with agitation at 37 °C. After 30, 60, and 90 min remove 300 μl aliquots. Centrifuge for 2 min in a microcentrifuge and transfer 200 μl of each supernatant to a fresh tube. Keep on ice until use.

Protocol 10. *Continued*

2. To each tube add 100 μl of a fresh overnight culture of the non-suppressor strain WK6. Incubate without agitation at 37 °C for 20 min and streak out on 2YT agar plates containing streptomycin (50 μg/ml) and ampicillin (100 μg/ml) or chloramphenicol (25 μg/ml), as appropriate. Incubate the plates overnight at 37 °C.

[a] BMH71–18/*mutS* strain genotype: Δ(*lac-pro*AB), *sup*E, *thi*, *mutS*::Tn 10; F' *lac*I^q, *lac*ZΔM15, *pro*A$^+$B$^+$.

[b] WK6 strain genotype: Δ(*lac-pro*AB), *thi*, *rpsL* (Str^r), *nal^r*; F' *lac*I^q, *lac*ZΔM15, *pro*A$^+$B$^+$.

[c] Chloramphenicol (1000 × stock solution) 25 mg/ml in 100% ethanol. Store at −20 °C.

[d] Ampicillin (1000 × stock solution) 100 mg/ml of ampicillin sodium salt in water. Sterilize by filtration (0.2 μm pore size) and store at −20 °C.

4. The deoxyuridine procedure

In this procedure, single-stranded DNA is isolated from a *dut, ung E. coli* strain. It contains a number of uracil residues instead of thymine. Note that uracil residues in the template DNA have the same coding potential as thymine. Thus, incorporation of U instead of T is non-mutagenic.

The mutagenic oligonucleotide serves as a primer for the complete synthesis of the complementary strand by native T7 DNA polymerase (a M_r 84 000 polypeptide derived from gene 5 of bacteriophage T7 and a M_r 12 000 polypetide, thioredoxin, of *E. coli*). It is advisable to check whether the filling-in reaction is both successful and complete by subjecting an aliquot to electrophoresis through a 1% TAE agarose gel containing 0.5 μg/ml ethidium bromide. As controls, single-stranded phasmid DNA, supercoiled covalently closed circular (ccc) DNA, and nicked circular DNA of the vector should be loaded in separate slots of the same gel. The *in vitro* polymerization product should migrate as supercoiled ccc DNA (*Protocol 11*).

Protocol 11. Directed mutagenesis using the deoxyuridine method

Equipment and reagents

- Competent cells of a Dut$^+$ *E. coli* strain, e.g. BMH71-18[a] (*Protocol 1*)
- Recombinant uracil-containing ss DNA (obtained from a *E. coli dut$^-$, ung$^-$* double mutant like BW313[b]; for preparation see *Protocol 4*)
- ccc DNA of the recombinant vector (approx. 100 ng)
- Nicked circular DNA of the recombinant vector (approx. 100 ng)
- 2YT agar plates (*Protocol 1*) containing the appropriate antibiotic
- 2YT medium (*Protocol 1*)

- 5′ Phosphorylated oligonucleotide (*Protocols 5–7*)
- T7 DNA polymerase form II unmodified (New England Biolabs)
- T4 DNA ligase (Boehringer-Mannheim Biochemicals)
- 10 × annealing buffer (*Protocol 9*)
- Dilution buffer (*Protocol 9*)
- 10 × synthesis buffer (*Protocol 9*)
- TAE buffer; 40 mM Tris–acetate, 2 mM EDTA, pH 8.0
- 1% (w/v) agarose in TAE buffer containing 5 μg/ml ethidium bromide

Method

1. Prepare the following reaction mixture in a microcentrifuge tube:
   - Uracil-containing single-stranded DNA
     (prepared according to *Protocol 4*)   0.3 pmol
   - phosphorylated oligonucleotide   10 pmol
   - 10 × annealing buffer   1 μl
   - H$_2$O   to 10 μl final volume

2. Prepare a second control reaction mixture containing all the above ingredients except the mutagenic primer.

3. Incubate for 5 min at 70 °C. Allow the reactions to cool to room temperature in the water-bath over a period of about 30 min. Transfer the reaction mixture into an ice/water-bath.

4. Dilute unmodified T7 DNA polymerase (the T7 gene 5 protein/thioredoxin complex) in dilution buffer to a final concentration of 1 U/μl.

5. Add to the reaction mixtures:
   - 10 × synthesis buffer   3 μl
   - H$_2$O   14 μl
   - T4 DNA ligase (1 U/μl)   2 μl
   - diluted T7 DNA polymerase (1 U/μl)   1 μl.

6. Incubate the mixtures sequentially on ice for 5 min, at 25 °C for 5 min, and finally at 37 °C for 60 min. Transfer the mixtures back to ice. If necessary, the procedure can be interrupted at this point by storing the reaction mixtures at −20 °C.

7. Analyse 10 μl of each reaction product by electrophoresis through a 1% agarose gel in TAE containing 0.5 μg/ml ethidium bromide. Apply 200 ng of the ss DNA, 100 ng ccc DNA, and 100 ng nicked circular DNA of the vector to separate lanes of the gel. A band migrating as supercoiled ccc DNA indicates successful conversion.

8. Prepare competent cells of a Dut$^+$, Ung$^+$ *E. coli* strain (e.g. BMH71-18) as described in *Protocol 1*. Add 10 μl of the reaction mixture to 200 μl of competent cells. Incubate on ice for 15 min. Transfer to 42 °C for 3 min, add 1 ml 2YT, and incubate at 37 °C for 60 min. Streak out 200 μl aliquots on selective plates (phasmid-borne antibiotic resistance).

[a] BMH71-18 strain genotype: Δ(*lac-proAB*), *supE*, *thi*; F' *lacI*^{q}, *lacZ*ΔM15, *proA*$^+$B$^+$.
[b] BW313 strain genotype: HfrKL16 PO/45 (*lysA61–62*), *dut1*, *ung1*, *thi1*, *relA1*.

Any Dut$^+$ *E. coli* strain may be used as transformation host for the heteroduplex DNA. Utilization of a mismatch repair deficient *E. coli* strain as transformation host has only a minor effect on mutagenesis efficiency (13). Marker yields routinely achieved by this method are between 50 and 90%.

5. The MHT procedure

5.1 General outline

The gapped duplex approach to oligonucleotide-directed mutation construction has been developed further, in a way that renders *in vitro* DNA polymerase/ligase reaction dispensable. The gapped duplex DNA can be introduced together with the mutagenic oligonucleotide directly into *E. coli*, where a polymerase/ligase reaction *in vivo* can rescue the sequence carried on the mutagenic primer into the newly synthesized strand (10, 20). Marker yields around 10 to 60% are achievable by using oligonucleotides that are protected against nucleolytic degradation *in vivo* by the incorporation of one or preferably two internucleotide phosphorothioate linkages directly adjacent to their 5′ terminus (20). Using this very simple mix-heat-transform (MHT) protocol, direct co-transfection of a mismatch repair deficient host strain with the respective gapped duplex DNA plus the mutagenic oligonucleotide is sufficient for obtaining the desired mutation in reasonable yields. This procedure can even be further simplified by omitting the gapped duplex DNA formation, hence by direct co-transformation of the mismatch repair deficient host bacteria with the mutagenic oligonucleotide and the single-stranded template DNA. With this procedure expectable marker yields are between 5 and 20%.

Besides the advantage of omitting the enzymatic reactions, this protocol, which leaves all the enzymatic reactions to the transformed cell, may yield less undesired nucleotide misincorporations in the course of the fill-in reaction *in vivo*.

5.2 Monitor of colony colour

In a variety of circumstances, it is useful to have a general macroscopic indicator at hand for the successful introduction of an oligonucleotide-encoded mutation. When structural genes are targets for mutagenesis, this can be achieved by cloning the target gene into a phasmid vector that contains the polylinker within the 5′ terminal coding sequence of the *lacZα* gene fragment, such as the pEMBL (26), pUC118/119 (21), or pMa/c series of vectors (11, 34). In a first round of mutagenesis, the target gene is fused in frame to the *lacZα* gene fragment and, simultaneously, an amber (TAG) stop codon is introduced between the two genes. Expression of the target gene either under control of its own or the *lac* promoter leads to production of a bipartite fusion protein in an amber suppressor host such as BMH71-18 and, consequently a blue colony phenotype on Xgal indicator plates.

Prior to the introduction of the desired mutation, a single nucleotide of the codon of interest in the target gene is removed by directed mutagenesis. This introduces a frameshift leading to a colourless colony on Xgal indicator plates. An altered codon can now be introduced concomitantly with restoration of the reading frame by using a synthetic oligonucleotide, which simultaneously brings about both changes. Thus, successful mutations are scored as blue colonies after transformation of a host strain carrying an amber suppressor. On the other hand, a non-suppressor strain may be used to produce the corresponding mutant protein.

In order to obtain pure progeny clones it is necessary to allow the mixed phasmid population resulting from the transformation of the heteroduplex DNA to segregate either via re-transfomation (*Protocol 10A*) or phasduction (*Protocol 10B*).

This procedure requires two successive rounds of mutagenesis to introduce the desired mutation. However, by using this means of alternatively detecting white or blue colony phenotypes, clones carrying the desired mutations can be immediately identified. The direct scoring of mutant clones by their colony phenotype alleviates the need of high marker yield and therefore enables this very simple mutagenesis procedure to be used successfully (see Section 5.1).

6. Problems and troubleshooting

6.1 DNA mismatch repair

When using a mutagenesis technique, in which the recipient cells are transformed with a heteroduplex DNA containing one or several base/base mismatches, it is generally advantageous to avoid unwelcome DNA mismatch repair processes which might work in favour of the template DNA strand. Hence, for the gapped duplex, deoxyuridine, and MHT procedures we recommend using an *E. coli mutHLS* strain, which exhibits a strongly reduced efficiency in the repair of many base/base mismatches. In rare cases however, other (*mutHLS* independent) repair pathways may also result in low marker yield. The VSP repair system is one such example. This recognizes a T/G mismatch in the sequence $CC^A/_TGG$ (36, 37). The *mutY* repair pathway (38), which is able to process G/A mismatches may also be a cause of low mutant yield. In such cases, the application of the phosphorothioate method, with which the transforming DNA is in a homoduplex state at the time of entering the recipient cell may be a helpful alternative.

6.2 DNA polymerase reaction *in vitro*

With the exception of the MHT procedure, all protocols described in this chapter require that a DNA polymerase reaction be carried out *in vitro*. Native T7 DNA polymerase has proven to be a highly processive enzyme (39), which is also able to copy DNA templates with intrinsic secondary structure. However, when using phasmid vectors that contain tandemly arranged strong transcription terminators, problems may arise in obtaining a full-length polymerization product. In these cases, the gapped duplex approach to oligonucleotide-directed mutation construction may be preferable, where the

DNA sequence with the tendency to form strong secondary structures can be entirely covered by the gapped strand.

Another cause of low mutant yield, especially when using the deoxuridine method, may be the contamination of the ss DNA preparation with small DNA or RNA molecules, which might serve as primers for the *in vitro* DNA polymerase reaction. Purity of the ss DNA can be checked by performing a control mutagenesis experiment with DNA polymerase/DNA ligase reaction but without the addition of the mutagenic oligonucleotide. A high yield of transformants indicates impurities in the single-stranded DNA preparation. In such cases, treatment of the ss DNA preparation with RNase A may be helpful. To this end, heat 5–10 μg ss DNA in 1 × annealing buffer (*Protocol 9*) together with 20 μg RNase A for 2 min to 100 °C and allow to cool to ambient temperature over a period of 30 min. RNA–DNA hybrids are not substrates for RNase A. Raising the temperature to 100 °C leads to the displacement of the RNA from the ss DNA template. This will liberate RNA enabling it to be digested by RNase A, which re-folds into its functional form upon lowering the incubation temperature.

6.3 Structure of the synthetic oligonucleotide

A major obstacle for successful incorporation of the mutagenic oligonucleotide into a phasmid vector is strand displacement, which can occur when DNA polymerase encounters a fraying 5′ end of the primer annealed to the template during the *in vitro* reaction. Whenever possible, the primer sequence at the 5′ terminus should therefore consist of one or several G or C residues.

Inverted repeats within the primer sequence gives the oligonucleotide self-complementarity and may therefore provoke oligonucleotide duplex formation. This may strongly interfere with its annealing to the target site. Direct repeats within the flanking region of the oligonucleotide can lead to primer–template slippage with loop formation on the primer or the template site. This could result in an undesired insertion or deletion.

In the cases mentioned above, the use of primers with longer flanking regions may be helpful. Correct hybridization of the mutagenic oligonucleotide to the target DNA can easily be checked by performing a nucleotide sequence analysis using the dideoxy method (40), where the mutagenic oligonucleotide serves as a sequencing primer.

References

1. Hutchison III, C. A. and Edgell, M. H. (1971). *J. Virol.*, **8**, 181.
2. Hutchison III, C. A., Phillips, S., Edgell, M. H., Gillam, S., Jahnke, P., and Smith, M. (1978) *J. Biol. Chem.* **253**, 6651.
3. Zoller, M. J. and Smith, M. (1983). In *Methods in enzymology* (ed. R. Wu), Vol. 100, pp. 468–500. Academic Press, London.

4. Winter, G., Fersht, A. R., Wilkinson, A. J., Zoller, M., and Smith, M. (1982). *Nature*, **299**, 756.

5. Kramer, W., Schughart, K., and Fritz, H.-J. (1982). *Nucleic Acids Res.*, **10**, 6475.

6. Kramer, W. and Fritz, H.-J. (1992). In *Modern methods in protein- and nucleic acid research* (ed. H. Tschesche), pp. 19–35. Walter de Gruyter, Berlin.

7. Sayers, J. R. and Eckstein, F. (1991). In *Directed mutagenesis: a practical approach* (ed. M. J. McPherson), pp. 83–99. IRL Press, Oxford.

8. Sayers, J. R., Krekel, C., and Eckstein, F. (1992). *BioTechniques*, **13**, 592.

9. Kramer, W., Drutsa, V., Jansen, H.-W., Kramer, B., Pfugfelder, M., and Fritz, H.-J. (1984). *Nucleic Acids Res.*, **12**, 9441.

10. Kramer, W. and Fritz, H.-J. (1987). In *Methods in enzymology* (ed. R. Wu), Vol. 154, pp. 350–67. Academic Press, London.

11. Stanssens, P., Opsomer, C., McKeown, Y. M., Kramer, W., Zabeau, M., and Fritz, H.-J. (1989). *Nucleic Acids Res.*, **17**, 4441.

12. Kramer, B., Kramer, W., and Fritz, H.-J. (1984). *Cell*, **38**, 879.

13. Kunkel, T. A., Roberts, J. D., and Zakour, R. A. (1987). In *Methods in enzymology* (ed. R. Wu), Vol. 154, pp. 367–82. Academic Press, London.

14. Kunkel, T. A., Bebenek, K., and McClary, J. (1990). In *Methods in enzymology* (ed. J. Miller), Vol. 204, pp. 125–39. Academic Press, London.

15. Bebenek, K. and Kunkel, T. A. (1989). *Nucleic Acids Res.*, **17**, 5408.

16. Lindahl, T. (1974). *Proc. Natl. Acad. Sci. USA*, **71**, 3649.

17. Lindahl, T. (1982). *Annu. Rev. Biochem.*, **51**, 61.

18. Konrad, E. B. and Lehmann, I. R. (1975). *Proc. Natl. Acad. Sci. USA*, **72**, 2150.

19. Tye, B.-K. and Lehmann, I. R. (1977). *J. Mol. Biol.*, **117**, 293.

20. Fritz, H.-J., Hohlmayer, J., Kramer, W., Ohmayer, A., and Wippler, J. (1988). *Nucleic Acids Res.*, **16**, 6987.

21. Vieira, J. and Messing, J. (1987). In *Methods in enzymology* (ed. R. Wu), Vol. 153, pp. 3–11. Academic Press, London.

22. Messing, J., Gronenborn, B., and Hofschneider, P. H. (1977). *Proc. Natl. Acad. Sci. USA*, **74**, 3642.

23. Messing, J. (1991). *Gene*, **100**, 3.

24. Messing, J. (1983). In *Methods in enzymology* (ed. R. Wu), Vol. 101, pp. 20–78. Academic Press, London.

25. Dente, L., Cesareni, G., and Cortese, R. (1983). *Nucleic Acids Res.*, **11**, 1645.

26. Dente, L. and Cortese, R. (1987). In *Methods in enzymology* (ed. R. Wu), Vol. 153, pp. 111–19. Academic Press, London.

27. Meyer, T. F., Geider, K., Kurz, C., and Schaller, H. (1979). *Nature*, **278**, 365.

28. Short, J. M., Fernandez, J. M., Sorge, J. A., and Huse, W. D. (1988). *Nucleic Acids Res.*, **16**, 7583.

29. Carter, P. (1991). In *Directed mutagenesis: a practical approach* (ed. M. J. McPherson), pp. 1–25. IRL Press, Oxford.

30. Devereux, J., Haeberli, P., and Smithies, O. (1984). *Nucleic Acids Res.*, **12**, 387.

31. Fritz, H.-J., Belagaje, R., Brown, E. L., Fritz, R. H., Jones, R. A., Lees, R. G., and Khorana, H. G. (1978). *Biochemistry*, **17**, 1257.

32. Wu, R., Wu, N.-H., Hanna, Z., Georges, F., and Narang, S. (1984). In *Oligonucleotide synthesis: a practical approach* (ed. M. J. Gait), pp. 135–51. IRL Press, Oxford.

33. Maniatis, T., Fritsch, E. F., and Sambrook, J. (ed.) (1982). *Molecular cloning, a laboratory manual*. Cold Spring Harbor Press, Cold Spring Harbor, NY.
34. Kolmar, H., Friedrich, K., Pschorr, J., and Fritz, H. -J. (1990). *Technique*, **2**, 237.
35. Friedrich, K., Kolmar, H., and Fritz, H.-J. (1989). *Nucleic Acids Res.*, **17**, 5862.
36. Lieb, M. (1983). *Mol. Gen. Genet.*, **191**, 118.
37. Lieb, M. (1985). *Mol. Gen. Genet.*, **199**, 465.
38. Au, K. G., Clark, S., Miller, J. H., and Modrich, P. (1989). *Proc. Natl. Acad. Sci. USA*, **86**, 8877.
39. Tabor, S., Huber, H. E., and Richardson, C. C. (1987). *J. Biol. Chem.*, **262**, 16212.
40. Sanger, F., Nicklen, S., and Coulson, A. R. (1977). *Proc. Natl. Acad. Sci. USA*, **74**, 5463.

7

DNA sequencing

LUKE ALPHEY

1. Introduction

There are two principal methods for determining the sequence of DNA. These are based either on specific chemical degradation of a DNA molecule labelled at one end (the Maxam and Gilbert method) or on the enzymatic synthesis of labelled DNA, using dideoxynucleotides to terminate the elongating strand (the Sanger dideoxy method). Although there are specific applications which require the Maxam and Gilbert method, the vast majority of DNA sequence is today determined by use of the dideoxy chain termination method.

2. Chemical degradation (Maxam and Gilbert)

The chemical degradation method (1) is based on the use of base-specific chemical cleavage of an end-labelled DNA molecule, followed by high resolution polyacrylamide gel electrophoresis to resolve the resulting fragments. The base-specific cleavage is achieved by using conditions in which a specific base or bases (i.e. G, G + A, C + T, or C) is modified in such a way that subsequent incubation with hot piperidine will cleave the sugar–phosphate backbone of the DNA at the sites of modification. The base-specific reactions are carefully designed to modify only a small proportion of the susceptible bases. Cleavage at these sites then yields a set of end-labelled molecules of one to several hundred nucleotides in length. Compared with the chain termination method, the chemical degradation method offers a number of advantages and disadvantages. Advantages relative to chain termination:

(a) Sequence is obtained from the original DNA molecule, rather than an enzymatic copy.

(b) Sequence can be determined very close to the labelled site (within two or three bases).

(c) Sequence can be determined from within DNA of unknown sequence, based only on the restriction map.

L. Alphey

Disadvantages relative to chain termination:

(a) More sequence can be determined from a single set of reactions by the chain termination method.

(b) Since the chemical degradation reactions all require different conditions, the chain termination method is generally faster and more reproducible.

As a consequence, the chemical degradation method is now used only where it is important to sequence the DNA directly, rather than via an enzymatic copy. This includes applications such as studies of DNA modification, secondary structure, and protein interaction, which are beyond the scope of this chapter (but see Chapter 5). For detailed protocols for sequencing by chemical degradation, see ref. 1.

3. Chain termination (Sanger dideoxy)

3.1 General introduction

The chain termination method (2) uses a DNA polymerase to extend a labelled DNA strand until a chain terminating 2',3'-dideoxynucleotide (ddNTP) is incorporated. These ddNTPs lack the 3'-OH group necessary for chain elongation. Four sets of these termination reactions are performed on each template, differing only in which of the four dideoxynucleotides is added. The resulting labelled strands form a nested set of specifically terminated chains up to several thousand bases long. These are separated according to size by high resolution denaturing gel electrophoresis.

3.2 Labelling methods

The synthesized strands may be labelled in any one of three different ways: at the 5' (primer) end, at the 3' (dideoxy) end, or throughout the length of the molecule. Of these, the last method is by far the most commonly used.

3.2.1 Labelling at the 5' (primer) end

In this method, the primer itself is labelled, for example by using T4 polynucleotide kinase and [γ-^{32}P]ATP or [γ-^{33}P]ATP. The subsequent extension/termination reaction uses only unlabelled nucleotides. This labelling method is commonly used for PCR-based sequencing protocols (see Section 3.10). The advantages of this method are:

(a) Only chains extended from the primer are labelled. This can be a significant advantage when the template is a large double-stranded molecule such as lambda.

(b) The single radiolabel present per chain results in an essentially uniform band intensity throughout the sequencing ladder.

(c) Radiolysis is not a problem as it simply results in unlabelled fragments.

226

End-labelled primers and reactions can be kept at $-20\,°C$ for up to one month before gel electrophoresis.

The disadvantages are:

(a) A separate kinase reaction is required to label the primer, which must be repeated regularly, and for each new primer.

(b) As only a single radiolabel is present per chain, the specific activity and hence band intensity is rather low. This means that isotopes with relatively low energy emissions such as ^{35}S cannot normally be used with this method (see Section 3.3).

(c) Not all γ-labelled ATPs are good substrates for the kinase: ^{35}S from [γ-^{35}S]ATPγS is incorporated much less efficiently than ^{32}P from [γ-^{32}P]ATP.

3.2.2 Labelling at the 3′ (dideoxy) end

This is used for some automated sequencing machines which incorporate coloured or fluorescent dyes at the 3′ end. These methods are beyond the scope of this chapter.

3.2.3 Labelling the synthesized chain

This can be performed in one or two steps, depending on the polymerase used. For Klenow and reverse transcriptases, the labelling and primer extension/chain termination reactions can be combined by lowering the concentration of one of the four dNTPs and adding the same radiolabelled dNTP. For all polymerases, including the widely-used T7 DNA polymerase, these two reactions can be performed sequentially. In the labelling reaction, the primer is extended a short distance using limiting concentrations of dNTPs and a single radiolabelled dNTP. In the extension/termination step the 'extended primers' are further extended in the presence of both dNTPs and ddNTPs, leading to sequence-specific chain terminations. The principle advantage of this method is that multiple radiolabels are incorporated into each chain, allowing the use of relatively low-energy β particle emitting isotopes such as ^{35}S.

3.3 Choice of radiolabel

The strands synthesized in a sequencing reaction are generally radiolabelled by one of the methods described above. The isotope used affects both the amount of sequence information that can be read from a single reaction and the exposure time required. The isotope of choice has traditionally been ^{35}S, whose relatively low energy β emission gives sharp bands on the autoradiogram, allowing longer and more accurate reading of the sequence. End-labelling methods give a lower specific activity for the synthesized strand and so must use more active isotopes such as ^{32}P. Unfortunately, the emission from ^{32}P is more penetrating, giving more diffuse bands and so less sequence information can be obtained from a given reaction. More recently, another isotope, ^{33}P, has become generally available. The characteristics of this isotope are intermediate between ^{32}P and ^{35}S and so this isotope is now sometimes used for end-labelling, and for labelling the synthesized strand in reactions which persistently yield faint bands on autoradiography when ^{35}S is used. [α-^{35}S]dATPαS is still used for most routine sequencing.

3.4 Choice of polymerase

The most commonly used polymerases for sequencing are the Klenow fragment of DNA polymerase 1, AMV reverse transcriptase, *Thermus aquaticus* (*Taq*) DNA polymerase, and T7 or modified T7 DNA polymerase (Sequenase, Sequenase v2.0). The Klenow fragment and reverse transcriptase have been used for many years, but are now largely superceded by the *Taq* and T7 enzymes. *Taq* DNA polymerase is thermostable and so the sequencing reactions can be run at a high temperature, minimizing the artefacts due to secondary structure. For most sequencing the enzyme of choice is Sequenase v2.0 T7 DNA polymerase (Amersham/USB US 70775, see Section 6.1), a modified T7 DNA polymerase in which the 3′–5′ exonuclease activity of the wild-type T7 enzyme has been removed, although I find little difference between Sequenase and T7 DNA polymerase (Pharmacia 27-0985-02). These enzymes show high speed, high processivity, and will readily incorporate base analogues such as ddNTPs and dITP, giving uniform band intensities and minimizing background bands.

3.5 Choice of primer

Many sequencing projects involve sequencing a set of clones of unknown sequence in a vector of known sequence. The primers used are complementary to vector sequence near the polylinker: most of the commonly used plasmid vectors have the M13 −20 and M13 reverse primer sites conveniently situated 20–40 nucleotides from the polylinker. These 'universal' primers are commercially available and have the sequences 5′ GTAAAACGACG-GCCAGT 3′ and 5′ AACAGCTATGACCATG 3′ respectively. In addition to these, custom oligonucleotides are widely-used for specific applications, such as sequencing different alleles of the same region, or completing the sequencing of cloned DNA without having to undertake exhaustive subcloning (see Section 6.3). Primers are typically 16–20 nucleotides long, but longer primers will also work.

3.6 Template preparation

While most nucleic acids can be sequenced in some way, the ideal template is single-stranded DNA, which can be obtained from suitable M13 or phagemid vectors. Many common plasmid vectors, e.g. pKS, pGEM-Zf, etc. include an f1 origin to allow them to be expressed as single-stranded DNA. Recombinant plasmids can also be sequenced directly as double-stranded DNA. This allows sequence information to be obtained from both ends of the insert, but the quality of sequence information obtained from double-stranded DNA is generally lower than from single-stranded DNA templates, and less can be read from each reaction.

3.6.1 Single-stranded DNA

Single-stranded DNA templates give the highest quality sequence and so I recommend this method for general purpose use. Phage particles containing single-stranded DNA can be prepared from plasmid vectors with a single-stranded origin (phagemids) by use of a helper phage (*Protocol 1A*) or M13 vectors can be used directly (*Protocol 1B*).

Protocol 1. Growing cultures for single-stranded DNA preparation

Reagents

- LB liquid media: 10 g Bacto tryptone, 5 g Bacto yeast extract, 10 g NaCl, make up to 1 litre with water, adjust to pH 7.0, and autoclave
- 100 mg/ml ampicillin stock solution, filter sterilized
- 100 mg/ml kanamycin stock solution, filter sterilized
- M13KO7 helper phage (e.g. Promega P2281)

A. *Phagemids*

1. Transform an appropriate *E. coli* host strain with double-stranded phagemid DNA (see text and Chapter 1).
2. Pick a single colony into 100 μl LB containing 50–100 μg/ml ampicillin in a microcentrifuge tube using a sterile micropipette tip.
3. Culture with agitation at 37 °C for 6–8 h.
4. Infect the culture with helper phage by mixing the following in a sterile 10–15 ml tube:
   - LB containing 50–100 μg/ml ampicillin   2.5 ml
   - bacterial culture from above (or from an overnight culture)   50 μl
   - M13KO7 helper phage (> 10^{11} p.f.u./ml)   3 μl
5. Shake the culture at 37 °C for 1 h.
6. Add kanamycin to 100 μg/ml.
7. Shake the culture at 37 °C overnight.

B. *M13 plaques*

1. Add 50 μl of a saturated culture of host cells to 2.5 ml LB.
2. Infect with M13 by touching a sterile toothpick to the plaque and washing it in the 2.5 ml culture.

L. Alphey

Protocol 1. *Continued*

3. Culture with agitation at 37 °C for 5–7 h. Growth for longer periods results in the accumulation of deletion derivatives of recombinant M13.

Template DNA can be prepared from the phage culture by PEG precipitation of the phage particles followed by extraction of the DNA from the phage (*Protocol 2*).

Protocol 2. Single-stranded template DNA from bacteriophage culture

Reagents

- 20% PEG 6000, 2.5 M NaCl
- TE: 10 mM Tris–HCl pH 8.0, 1 mM EDTA
- TE-saturated phenol
- Chloroform (CHCl$_3$)
- 5 M ammonium acetate
- Ethanol

A. *Purifying phage particles from the culture*

1. From the 2.5 ml culture from *Protocol 1*, transfer 1.5 ml to a microcentrifuge tube. Centrifuge at 12 000 *g* for 5 min. Transfer 1.3 µl of the supernatant to a fresh tube, avoiding the bacterial pellet.

2. Add 200 µl of 20% PEG 6000, 2.5 M NaCl, mix well, and allow precipitate to form for 5 min at room temperature.

3. Centrifuge at 12 000 *g* for 5 min in a microcentrifuge. Remove and discard supernatant by careful aspiration.

4. Centrifuge at 12 000 *g* for 30 sec. Carefully remove any remaining PEG solution. Residual PEG can severely affect the subsequent sequencing reactions.

5. Resuspend the pellet of phage particles in 100 µl TE by vigorous vortexing. The phage suspension is stable at 4 °C indefinitely.

B. *Purifying single-stranded DNA from phage particles*

1. To the 100 µl of phage suspension add 100 µl TE-saturated phenol, vortex for 30 sec, and centrifuge at 12 000 *g* for 5 min in a microcentrifuge. Transfer the supernatant to a fresh tube, carefully avoiding the interface.

2. Add 100 µl chloroform (CHCl$_3$) to this supernatant, vortex, and centrifuge as above. Again transfer the supernatant to a fresh tube, avoiding the small amount of white precipitate generally seen at the interface.

3. Ethanol precipitate by adding:
   - 5 M ammonium acetate   50 µl
   - ethanol   350 µl.

   Place on dry ice for 30 min or at −20 °C overnight.

4. Centrifuge at 12 000 *g* for 10 min and carefully remove the supernatant. The pellet of single-stranded DNA is often streaked up the side of the tube and may not be visible. Air dry or vacuum dry the pellet. It is not necessary to completely remove all the ethanol and it is unwise to risk losing the pellet by trying to rinse it.

5. Resuspend the DNA in 20–50 μl of TE or water and store at −20 °C. The quality and quantity of the DNA can be checked by subjecting 2–5 μl to agarose gel electrophoresis. Note that single-stranded DNA has a much higher mobility on such gels than the equivalent double-stranded DNA.

3.6.2 Double-stranded DNA

Double-stranded DNA templates must be pure and RNA-free. Plasmid DNA purified on a CsCl gradient is suitable, as are various proprietary methods, e.g. Qiagen preps (Qiagen Cat. No. 12145) and Magic/Wizard DNA preps (Promega Cat. No. 7300). Problems arise when sequencing DNA prepared as a 'mini-prep', using for example the alkaline lysis method (3). We have found Magic/Wizard Minipreps (Promega A7500) to be generally satisfactory. Other methods such as the use of PEG precipitation (3) seem to work for some workers and not for others. Double-stranded DNA templates do not give good sequence as consistently as single-stranded templates, and cannot normally be read as far. Co-termination ('pausing') can be a problem with all but the cleanest templates. Larger templates such as lambda phage and cosmids do not often give satisfactory sequence with normal methods, although it is possible to sequence from such templates using *Taq* DNA polymerase at 70 °C and cycle sequencing (see Sections 3.9 and 3.10).

3.7 Annealing the primer

Protocols for annealing the primer to the template are given in *Protocol 3*. In the case of single-stranded templates this is very straightforward as the template does not need denaturing, whereas for double-stranded DNA the template must be denatured and not allowed to re-anneal before the primer is added. The nominal stoichiometry of primer to template is 1:1, but an excess of primer of up to tenfold makes little difference. Insufficient quantity of primer or template narrows the effective sequencing range, resulting in faint bands near the primer (see Section 5.3).

Protocol 3. The annealing reaction

Reagents

- 5 × annealing buffer: 200 mM Tris–HCl pH 7.5, 100 mM $MgCl_2$, 250 mM NaCl
- Sequencing primer (0.5–1.0 μM)
- 5 M NaOH
- 0.5 M EDTA
- 3 M sodium acetate pH 4.5–5.5
- TES buffer: 560 mM TES (free acid, e.g. Sigma Cat. No. T-4152), 240 mM HCl, 100 mM $MgCl_2$

Protocol 3. *Continued*

A single annealing reaction is used for each template and subsequently divided into four after the annealing reaction. Use method A for a single-stranded DNA template and either method B or method C for a double-stranded template.

A. *Single-stranded DNA template*

1. To an Eppendorf tube add the following:
   - single-stranded DNA template (*Protocol 2*)   7 μl
   - 5 × annealing buffer   2 μl
   - primer (0.5–1.0 μM)   1 μl.

2. Heat a beaker of water to 65–70 °C and immerse the reaction tubes. Remove from the heat and allow to cool slowly to below 30 °C, at which point the annealing reaction is completed. This cooling should take 15–30 min. The templates can be used immediately for the labelling reaction or stored overnight at −20 °C.

B. *Double-stranded DNA template*

1. Use at least 1 μg and preferably 3 μg DNA in 20–50 μl. Denature the DNA by adding:
   - NaOH to 0.2 M
   - EDTA to 0.2 mM.

   Incubate at 37 °C for 30 min.

2. Neutralize and precipitate by adding:
   - 3 M sodium acetate pH 4.5–5.5   0.1 vol.
   - ethanol   2–4 vol.

   Incubate on dry ice or at −70 °C for at least 15 min. Centrifuge at 12 000 g in a microcentrifuge for 5 min. Remove and discard the ethanol supernatant. Rinse the pellet briefly with 70% ethanol, remove, and air dry. It is not necessary to dry the pellet thoroughly.

3. Resuspend and anneal the pelleted DNA by adding:
   - distilled water   7 μl
   - annealing buffer   2 μl
   - primer (0.5–1.0 μM)   1 μl.

   Vortex briefly to resuspend the DNA and incubate at 37 °C for 15–30 min. The annealed DNA can be used immediately or stored overnight at −20 °C.

C. *Alkaline denaturation and TES neutralization*

This method relies on the accurate neutralization of the NaOH by the addition of TES buffer. The exact amount of TES buffer required to bring the sequencing mix to pH 7.0 must be checked for each batch of buffer and the protocol adjusted accordingly. Accurate pipetting of the NaOH and TES is required.

1. Denature the DNA by mixing the following:
   - DNA (3 μg at 0.5 μg/μl)   6 μl
   - primer (0.5–1.0 μM)   1 μl
   - freshly diluted 1 M NaOH   1 μl.

   Incubate at 68 °C for 10 min.

2. Remove the tube containing the DNA from the water-bath and quickly add 1–2 μl TES buffer to neutralize the solution (see above). Leave for 10 min at room temperature. Centrifuge briefly to bring down condensation. The annealed DNA is now ready for the sequencing reactions.

3.8 The sequencing reactions

Efficient and reproducible sequencing requires careful preparation to ensure that all reagents are on hand as there is considerable time pressure on the operator if a number of templates need sequencing. I sequence annealed templates in batches of ten, with four to six batches taking most of an afternoon. I find that it is much faster to use a microtitre dish for the sequencing reactions, rather than sets of microcentrifuge tubes. There are a wide-range of microtitre dishes available. I use Greiner Labortechnik, Cat No. 653180. These have a well volume of about 15 μl, and seem to allow the reactions to get to temperature quickly.

Do not use microtitre dishes for the annealing reaction as evaporation will seriously alter the concentration and volume.

Thaw out all reagents that you will need, except the polymerase, and keep on ice. Make up a mix of sufficient diluted dNTP concentrate, DTT, and [α-^{35}S]dATP for your reactions.

Chill a microtitre plate on ice and add termination mixes to wells in an appropriate pattern, e.g. rows of G, A, T, C containing ddGTP termination mix, ddATP, etc.

Add the diluted polymerase to the labelling reaction mix and start the labelling reactions at 20–30 second intervals by removing an annealed template from ice and adding 5 μl of the labelling reaction mix.

One to two minutes before starting the termination reactions, place the microtitre dish with the termination mixes on to a 37–45 °C heating block. Wetting the surface of the block improves the thermal contact. A water-bath can be used instead, but is much less convenient.

Pipetting out the labelling reaction into each of the four termination mixes will take approximately 20–30 seconds per template, as will adding the stop solution, so each reaction gets approximately the same time at each step.

Protocol 4. The sequencing reactions [a]

Equipment and reagents

- Heating block (or water-bath)
- Microtitre dish (see text)
- 5 × dNTP concentrate: 7.5 μM dCTP, 7.5 μM dGTP, 7.5 μM dTTP
- 0.1 M DTT (dithiothreitol)
- [α-^{35}S]dATPαS, e.g. Amersham SJ264, SJ304, or SJ1304 (15–37 TBq/mmol, 370 MBq/ml)
- T7 DNA polymerase (7000 U/ml, see text)

- Polymerase dilution buffer: 10 mM Tris–HCl pH 7.5, 5 mM DTT, 0.5 mg/ml BSA
- Termination mixes: core solution is 80 μM of each dNTP + 50 mM NaCl (add 8 μM ddGTP for ddG mix, 8 μM ddATP for ddA mix, etc.)
- Stop solution: 95% formamide, 20 mM EDTA pH 8.0, 0.05% bromophenol blue, 0.05% xylene cyanol FF

A. *The labelling reaction*

1. To the annealed template and primer add the following on ice:
   - template–primer from annealing reaction   10 μl
   - dNTP concentrate diluted 5 × in water   2.0 μl
   - 0.1 M DTT   1.0 μl
   - [α-^{35}S]dATPαS   0.1–0.5 μl
   - T7 DNA polymerase diluted 10 ×   2.0 μl.

   Mix by pipetting, avoiding bubbles. The polymerase is diluted in ice-cold polymerase dilution buffer. Diluted enzyme is stable for only about 30 min.

2. Incubate for 2–5 min at room temperature. As the reaction has only just been removed from ice, the actual temperature will be rather less than room temperature, which minimizes sequencing artefacts near the primer.

B. *The termination reactions*

1. Before adding the DNA polymerase to start the labelling reactions, prepare a microtitre dish by labelling four columns of wells G, A, T, and C respectively. Add 2.5 μl of ddGTP termination mix to the G wells, ddATP mix to the A wells, etc. and keep on ice.

2. Pre-warm the dish to 37–45 °C for at least 1 min, to be ready for the end of the labelling reaction.

3. When the labelling reaction is complete, pipette 3.5 μl of it into one well each of G, A, T, and C, and mix by pipetting, avoiding bubbles. Incubate at 37–45 °C for 3–5 min.

4. Add 4 μl of stop solution to each reaction and store at −20 °C until ready to load the sequencing gel. Reactions are stable for about a week.

[a] The modifications to this protocol required for ITP sequencing are described in *Protocol 7.*

3.9 Sequencing bacteriophage DNA

Linear double-stranded DNA can be converted to a single-stranded form by the use of bacteriophage T7 gene 6 exonuclease (Amersham/USB Cat. No. US 70025). This enzyme hydrolyses double-stranded DNA from 5′ termini. The products of this reaction are half-length, single-stranded DNA molecules. These can be used as single-stranded templates in *Protocol 3A*. This method can be applied to any double-stranded DNA, and may also be useful for the direct sequencing of PCR products.

3.10 Cycle sequencing

Cycle sequencing uses a thermal cycler and a heat-stable DNA polymerase to anneal the primer to the template, perform the extension (sequencing) re-action at an elevated temperature, then separate the labelled strand from the template by heat denaturation, and repeat the process as many times as required. Compared to conventional dideoxy sequencing (*Protocols 3 and 4*), this method offers various advantages and disadvantages.

Advantages of cycle sequencing:

(a) Allows the use of much lower levels of template DNA, typically 50 fmol (the mass of 50 fmol of double-stranded DNA is approximately 33 ng/kb).

(b) Elevated reaction temperature reduces artefacts due to template second-ary structure and routine use of 7-deaza-dGTP reduces compression artefacts.

(c) Only sequences derived from the end-labelled primer will be detected. This method is less sensitive to template impurities.

Disadvantages of cycle sequencing:

(a) Requires the use of ^{32}P or ^{33}P rather than ^{35}S, as neither T4 DNA kinase, nor *Taq* polymerase will efficiently utilize thionucleotides. The relative merits of these isotopes are discussed in Section 3.2.1.

(b) Sequence data quality is lower than for conventional sequencing, and so less information can obtained from each reaction.

(c) Cycle sequencing is less consistent and reliable than conventional sequencing.

As a consequence, cycle sequencing is only used when the amount of DNA available is limiting. This includes sequencing products of the polymerase chain reaction (PCR), and also λ phage and cosmid clones, where the large size of the clone limits the molar concentration in the sequencing reaction. Direct sequencing of PCR products is faster than subcloning and sequencing and also has the advantage that the average population of product molecules is sequenced, rather than a single one, which minimizes errors due to the low fidelity of *Taq* polymerase.

Stratagene's Cyclist Exo⁻ *Pfu* DNA sequencing kit (Cat. No. 200326) is an exception to the above generalizations. This system uses a mutagenized version of a cloned *Pyrococcus furiosus* DNA polymerase. Unlike *Taq* polymerase, *Pfu* polymerase will incorporate $[\alpha\text{-}^{35}S]$dATPαS efficiently. This removes the need for end-labelling. However, this also removes the advantage of end-labelling that only strands synthesized from the sequencing primer will be detected, and so this system is not suitable for direct sequencing of DNA from plaques or colonies (*Protocol 6*) or with lambda or cosmid templates. This system is only likely to be useful for direct sequencing of PCR products.

3.10.1 Template preparation for cycle sequencing

PCR products must be purified before sequencing. Nucleotides and primers carried over from the amplification will interfere with the sequencing reactions. It may also be necessary to separate multiple products by agarose gel electrophoresis prior to sequencing, particularly if one of the PCR primers is to be used as the sequencing primer. PCR products can be purified by use of commercially available kits, e.g. Magic/Wizard PCR Cleanup (Promega A7170) or QIAquick (Qiagen Cat. No. 28104), or simply by propan-2-ol precipitation as described in *Protocol 5*.

Protocol 5. Purification of PCR amplified DNA

Reagents

- TE: 10 mM Tris–HCl pH 7.5, 1 mM EDTA
- 200 fmol amplified DNA in TE or water[a]
- 4 M ammonium acetate
- Propan-2-ol (isopropanol)

Method

1. Dilute DNA solution to 20 μl in a microcentrifuge tube, add:
   - 20 μl 4 M ammonium acetate
   - 40 μl propan-2-ol.

   Mix well, allow to precipitate for 10 min at room temperature.
2. Centrifuge at 12 000 *g* for 10 min. Discard supernatant, carefully rinse pellet with 70% ethanol. Remove ethanol and allow to air dry.
3. Resuspend pellet in 20 μl TE. Use 5 μl (50 fmol) as the template in *Protocol 7*.

[a] The mass of 50 fmol of double-stranded DNA is approximately 33 ng/kb.

As very little template DNA is required for cycle sequencing, it is possible to extract DNA directly from a single colony or plaque and use this as the template. Various lysis methods are suitable, but avoid the use of detergents as these may inhibit the DNA polymerase. A simple lysis method is given in *Protocol 6*.

Protocol 6. Template DNA for cycle sequencing from a single colony or plaque

Reagents

• Lysis solution: 10 mM Tris–HCl pH 7.5, 1 mM EDTA, 50 μg/ml proteinase K (e.g. Gibco-BRL 540-5530UA)

Method

1. (a) For a single colony scrape a large (1–2 mm diameter) colony from an agar plate. Transfer to 12 μl of lysis solution in a microcentrifuge tube.
   (b) For a single plaque (M13 or lambda)[a] punch through agar around plaque (leave the plug in the plate). With a sterile toothpick or scalpel blade, carefully remove the top agar containing the plaque and transfer to 12 μl of lysis solution in a microcentrifuge tube.[b]

2. Vortex 30 sec.

3. Incubate the tube at 55°C for 15 min.

4. Incubate the tube at 80°C for 15 min to inactivate the protease.

5. Cool on ice for 1 min. Spin at 12000 g in a microcentrifuge for 5 min. Carefully remove 9 μl of supernatant from the surface. Use the entire 9 μl as the template in *Protocol 8*.

---

[a] Various lambda strains give different plaque sizes containing variable amounts of λ DNA. A fresh, large (2 mm) plaque is best, smaller or older plaques yield less DNA and fainter sequence bands.
[b] Contaminants in the agar may interfere with the subsequent reactions, so take care to minimize the carry-over of agar.

3.10.2 Cycle sequencing reactions

A range of cycle sequencing kits are commercially available (e.g. Gibco-BRL Cat. No. 819SA/SB, Amersham/USB US 71075). As with conventional sequencing, I advise buying a kit first, and then supplementing this with your own reagents as the kit runs out (see Section 6.1). *Protocol 7* is designed for use with *Taq* polymerase. Other polymerases may incorporate nucleotide analogues with different efficiencies, requiring modification of the termination mixes, or may be more or less thermostable. You will need a primer of at least 20 bases, labelled at the 5′ end using T4 DNA kinase. A protocol for end-labelling is given in Chapter 4, *Protocol 4*. No further purification of the oligo is required after this labelling reaction. As the desired PCR product is often contaminated with other DNAs that include the PCR primers, it is safer not to use either of the PCR primers as the sequencing primer.

Protocol 7. Cycle sequencing

Equipment and reagents

- Thermal cycler (e.g. Perkin-Elmer Cetus N801–0150 or Techne PHC-2)
- 5′ Labelled sequencing primer (20+ bases, 1 pmol/template)
- 10 × *Taq* polymerase reaction buffer: 300 mM Tris–HCl pH 9.5, 50 mM MgCl$_2$, 300 mM KCl, 0.5% (w/v) Tween-20 or NP-40
- *Taq* DNA polymerase (5 U/μl) (e.g. Amersham/USB US71057)

- Termination mixes: core mix is 50 μM each of dATP, dCTP, 7-deaza-dGTP, and dTTP; to this core mix add: 2 mM ddATP (ddA mix), or 1 mM ddCTP (ddC mix), or 0.2 mM ddGTP (ddG mix), or 2 mM ddTTP (ddT mix)
- Mineral oil (e.g. BDH 14017)
- Stop solution: 95% formamide, 20 mM EDTA pH 8.0, 0.05% bromophenol blue, 0.05% xylene cyanol FF

Method

1. Thaw each reaction component, and store on ice. Label four microcentrifuge tubes[a] 'A', 'C', 'G', and 'T'. Add 2 μl of termination mix A to tube 'A'. Similarly, add 2 μl of mix C to tube 'C', etc. Store these tubes on ice.

2. To a clean, ice-cold microcentrifuge tube add:
   - 10 × *Taq* polymerase reaction buffer   4.5 μl
   - template DNA (~ 50 fmol) + water   26 μl
   - labelled primer (1 pmol)   5 μl
   - *Taq* DNA polymerase (5 U/μl)   0.5 μl.

   Mix gently, centrifuge briefly to collect the mix at the bottom of the tube.

3. Add 8 μl of the mix from step **2** to each of the tubes from step **1**. Overlay each reaction mix with 10–20 μl of mineral oil.

4. Pre-heat the thermal cycler to 95 °C. Place reaction tubes in thermal cycler. Start cycler program. Appropriate programs depend on the application. Purified DNA, e.g. PCR products, purified lambda, and cosmid DNAs:
   30 cycles:
   - 30 sec at 95 °C
   - 30 sec at 55 °C[b]
   - 60 sec at 70 °C.

   Unpurified DNA, e.g. from *Protocol 6*:
   30 cycles:
   - 1 min at 95 °C
   - 1 min at 55 °C[b]
   - 2 min at 70 °C.

   Remove tubes from thermal cycler.

5. (Optional.) Add 50 μl chloroform to each reaction, spin briefly in a microcentrifuge to separate the phases. Take the top phase by pipet-ting, carefully avoiding the bottom (organic) phase.

6. Terminate reactions by adding 5 μl stop solution to each tube. Store at −20 °C until ready to run on a gel (*Protocol 9*). The reaction products should be run on a gel as soon as possible (within 24 h).

---

[a] These microcentrifuge tubes must fit into your thermal cycler. 1.5 ml or 0.5 ml tubes are equally satisfactory. For larger numbers of reactions, a multiwell microtitre plate will be prefer-able. Again, this must be a heat-resistant one that fits your thermal cycler.

[b] The optimal annealing temperature will depend on the length and base composition of your primer. See Chapter 5, Section 2.4.2.

3.11 Non-radioactive methods

All the methods discussed so far rely on the incorporation of a radiolabel into the synthesized strand. Although this is by far the most commonly used approach, there are a range of non-radioactive alternatives available. Auto-mated sequencing machines (ABI 373A) use fluorescently labelled primers or dideoxynucleotides and sophisticated detection equipment, but at a high price. Most other methods require the transfer of the DNA from the sequencing gel to a solid support, e.g. a nylon membrane. The DNA in this case has been labelled not with a radionucleotide, but with some other modified nucleotide, e.g. with biotin (Amersham/USB US71350) or with digoxygenin (Boehringer Cat. No. 1449 443). These are then detected using a chemiluminescent detec-tion system and the results recorded on X-ray film, as for conventional sequencing. An exception is Promega's Silver Sequence system (Promega Q4130) in which the gel is stained directly to reveal the bands. A hard copy can be made on to a special film (Kodak EDF), which will be essential for most *de novo* sequencing projects where old results need to be re-examined to resolve discrepancies (see Sections 6.3 and 6.4).

The advantages of these systems are:

(a) They do not require the use of radioisotopes.

(b) The time from running the gel to getting the result is generally reduced as the 12–24 h autoradiography step is eliminated.

The disadvantages are:

(a) The data obtained is rarely of such high quality as that obtained by autoradiography.

(b) The 'hands-on' time is generally increased by the additional blotting and washing steps required.

At the present time, non-radioactive methods other than automated sequencers are rarely used.

4. Gel electrophoresis

Gel electrophoresis is often the limiting step that determines how much sequence can be read from a given set of reactions. The amount of sequence that can be read, and the ease with which it is obtained, depend on both the gel system used and the way in which the electrophoresis is performed.

4.1 Gel systems

The wide range of sequencing gel apparatus available makes it difficult to offer general advice. Each gel system comes with protocols for setting-up and running the gel. Some specific problems are covered in the next section. Here I will describe the general parameters involved.

(a) Gel plates are normally made of glass. As they need to be handled a lot for washing and then for pouring and running the gel, they will invariably get chipped and break. This can of course be minimized by careful handling and storage but bear in mind that they will be cheaper to replace in a system that uses a simple design, rather than one which has plates cast to a more complicated shape.

(b) Sequencing gels are typically 40–50 cm from top to bottom. This is a balance between the better resolution of longer gels and the difficulty in handling longer gels. Gels up to at least 100 cm can be run on some sets of apparatus, but fixing, drying, and exposing these gels becomes more difficult.

(c) Narrow gels are easier to handle, but I prefer a wide gel (30–40 cm) which allows nearly 100 lanes to be run on a single gel. These wide gels do not work so well with buffer gradient gels.

(d) Gels are typically 0.4 mm thick. This is a balance between the difficulty in handling thinner gels and their better resolution. The thickness depends on the spacers. Wedge gels are not recommended. While they reduce the voltage gradient at the bottom of the gel, and so give a more even spatial distribution of the bands than conventional gels, they take much longer to dry than normal gels and do not seem to resolve the bands so well.

(e) Sharks-tooth combs give more wells in the gel than conventional combs and allow more sequence to be read as the bands are actually adjacent to each other. With practice, they are better in every respect. Make sure that the combs are the same thickness as the spacers.

(f) A gel apparatus needs some system for dissipating the temperature gradient that builds up across a gel, or the samples will not run evenly. This can be a metal plate attached to the back of the gel plates, other gel systems use the buffer reservoir as a heat sink.

4.2 Making and running the gel

In order to get the best quality sequence it is essential to use reagents of the highest possible quality for the gel. For safety, reproducibility, and convenience liquid acrylamide mixes designed for sequencing are recommended (e.g. Sequagel, National Diagnostics EC830/835/840). If solid acrylamide is used, make up solutions fresh at least once a week and store in the dark.

Protocol 8. Making a sequencing gel

Equipment and reagents

- Sequencing gel apparatus, e.g. BRL model S2[a]
- 70% ethanol
- Siliconizing solution (e.g. Repelcote, BDH 6316 4J)
- 4 cm adhesive tape
- Bulldog clips
- Ammonium persulfate, 10% solution in water (keeps up to one week at 4°C)
- TEMED
- 10 × TBE pH 8.8: 540 g Tris base, 275 g, boric acid, 46.5 g sodium EDTA, distilled water to 1 litre

If using commercial acrylamide/urea solutions:
- Acrylamide concentrate[b]
- Diluent
- 10 × TBE for gel

If using 'home-made' solutions:
- 40% acrylamide solution[b]: 380 g acrylamide (sequencing grade), 20 g N'-methylenebisacrylamide, distilled water to 1 litre—warm to 37°C until dissolved (store at room temperature in the dark)
- Urea

A. *Preparation*

1. Clean glass plates thoroughly by washing sequentially with hot water and domestic detergent, distilled water, and 70% ethanol. Allow to air dry. In subsequent manipulations, be careful not to touch the clean side of the plates.

2. Siliconize one of the plates, e.g. the smaller one, by pipetting 300–500 μl of siliconizing solution on to the clean side and spreading it with a dry paper towel. Keep polishing the whole surface of the plate until the solution has evaporated. Make sure that everyone in the laboratory always uses the same convention or both plates will end up siliconized.

B. *Setting up the gel plates*

1. Assemble the gel plates with spacers at sides. Clean the combs ready for use. Seal around the sides and bottom edge with 4 cm adhesive tape, paying special attention to the corners. In some systems, applying a very small amount of petroleum gel to the spacers gives a better seal, but this should not normally be necessary.

2. Clamp the edges together with bulldog clips positioned directly over the spacers. Prop up the plates so that there is a slope of about 30° from top to bottom and a slight slope from one side to the other.

241

L. Alphey

Protocol 8. *Continued*

3. Make a reservoir into which the gel mix will be poured by taping across from one spacer 'ear'.

C. *Pouring the gel*

1. Cover the whole working area with absorbent paper as it is nearly impossible to pour a gel without dripping acrylamide solution on to the bench. For safety and convenience I prefer to use liquid acrylamide solutions which have already been made up with urea and a stabilizer (e.g. Sequagel, National Diagnostics EC830/835/840). Making up the gel solution with these pre-prepared solutions is easier and will be described first. My sequencing gels take 60 ml of gel solution so I make up 75 ml for each gel to be sure of having enough after leakage and spills. The following are for 100 ml of gel mix.

2. (a) Using Sequagel acrylamide/urea concentrates to make a 5% gel mix together the following:
   - buffer (10 × TBE)   10 ml
   - acrylamide concentrate   25 ml
   - diluent   65 ml
   - ammonium persulfate (10%)   1.0 ml
   - TEMED   20 µl.

   Try to avoid bubbles.
   (b) Using separate acrylamide and urea solutions to make a 5% gel mix together the following:
   - 40% acrylamide solution   12.5 ml
   - urea   42 g
   - 10 × TBE (pH 8.8)   10 ml
   - distilled water to   99 ml[c]
   - ammonium persulfate (10%)   1.0 ml
   - TEMED   20 µl.

3. Slowly pour the gel mix into the gel mould. The risk of bubbles in the gel will be minimized by pouring slowly and continuously rather than intermittently, and by using scrupulously clean gel plates. When the gel is nearly full, stop pouring and lie the gel down almost horizontal. This reduces its capacity so that the gel mix should now be overflowing at the top. Insert the flat side of the sharks-tooth combs about 3 mm into the mould, clamp into place to ensure a tight fit of the comb after polymerization, and leave to set.

4. The gel usually sets in about 30 min, but polymerization continues after this and gels are best left overnight, so they can be prepared the day before use. After polymerization, lay damp paper towels on the top of

the gel and wrap with cling-film, to prevent the top of the gel from drying out overnight.

[a] Some manufacturers' gel systems have their own gel casting apparatus and methods. The method described here is based on the BRL model S2, a typical system with two glass plates, 0.4 mm spacers, and sharks-tooth combs. Volumes of gel mix etc. should be adjusted for other systems.

[b] Acrylamide is a neurotoxin, so wear gloves when handling gel solutions.

[c] Sequencing gels prepared with 'home-made' solutions generally need degassing as dissolved gasses inhibit polymerization. Commercial preparations have additives to overcome this. Degas before adding ammonium persulfate or TEMED by applying low vacuum, e.g. from a water pump, until the solution stops emitting gasses, i.e. stops bubbling.

Protocol 9. Running a sequencing gel

Equipment and reagents

- Sequencing gel apparatus and polymerized gel from *Protocol 8*
- High voltage power pack (e.g. Pharmacia/ LKB 3000/150)
- Gel dryer (e.g. Genetic Research Instrumentation AE.3750)
- Vacuum pump
- 1 × TBE (*Protocol 8*)

- 10% methanol/10% acetic acid
- Large sheet of filter paper (e.g. Whatman 3030 917)
- X-ray film (e.g. Kodak X-Omat LS Cat. No. 181 1769)
- X-ray film cassettes (no intensifying screen) (e.g. Genetic Research Instrumentation 400/ RG10)

Method

1. Starting with the polymerized gel from *Protocol 8*, remove cling-film and paper towels. Wash under tap to remove unpolymerized acrylamide. Remove the tape and bulldog clips from the gel sandwich. Assemble the gel electrophoresis apparatus according to the manufacturers' instructions. This will typically involve clamping the gel sandwich to a back plate, creating the top buffer tank. Fill the top and bottom buffer tanks with sufficient 1 × TBE to make contact with the electrodes and the gel. Check that no buffer is leaking down from the top tank. Carefully remove the sharks-tooth combs, clean if necessary, turn over, and replace so that the teeth just touch the gel.

2. Pre-run the gel for 15–60 min before loading the samples. Sequencing gels are run at constant power. For the BRL S2 this is 70–100 W. The resistance of the gel changes during the run, so the voltage required is initially about 1500 V, rising to 2000 V.

3. Use a Pasteur pipette to wash out the wells carefully just before loading. Urea diffuses out of the gel and interferes with the loading and running of the samples if it is not removed by washing.

4. Denature the samples by heating to 85–95 °C for 2–5 min in a waterbath. Cool rapidly on ice to prevent re-annealing.

243

Protocol 9. *Continued*

5. Load the gel with the four reactions in adjacent lanes, e.g. GATC. Always load the lanes in the same order. If there is space on the gel, then loading each reaction twice in the order GATCTGCA ensures that each lane is adjacent to every other lane, which will make reading the sequence easier. Load 0.5–1.5 µl/lane. Special tips with fine, flattened ends are available for loading sequencing gels (e.g. Stratagene 410004). Alternatively use a standard yellow tip (e.g. Stratagene 410116), modified by flattening the tip under a scalpel handle. The same tip can be reused for the whole gel as the resulting contamination from one lane to the next is negligible.

6. Run the gel at the same power as for pre-running (step **2** above). Gels are run for 3–10 h or more. As a rough guide, in a 5% gel the bromo-phenol blue dye co-migrates with 35 base fragments and the xylene cyanol with 130 base fragments.

7. After running, disassemble the gel apparatus. Remember that the bottom buffer contains ^{35}S and dispose of it accordingly. Remove the combs and spacers. Separate the plates — the gel should stick to the non-siliconized plate. In a large tray, fix the gel for 15–30 min in 10% acetic acid/10% methanol to remove the urea. This improves the re-solution and speeds the drying. The times are for 0.4 mm gels, for gels of different thickness fix for proportionally longer or shorter. Wedge gels need at least 60 min.

8. Dry ^{35}S gels directly on to a sheet of filter paper (e.g. Whatman 3030 917). Cover with cling-film while drying, but remove this before ex-posing to X-ray film. Make sure that the gel is pressed flat against the film in the film cassette. 24–48 h exposures are usually sufficient.

4.3 Glycerol tolerant gels

High concentrations of glycerol in the sequencing reaction increase the stability of the T7 DNA polymerase. This allows the reaction to be per-formed at a higher temperature, with the potential elimination of some secondary structure effects. Replace the polymerase dilution buffer in *Protocol 4* with the same buffer containing 50% glycerol. This increases the glycerol concentration in the sequencing reaction from 0.8% to 6%. This glycerol concentration can be further increased up to 50% if required. However, glycerol interferes with the gel electrophoresis, distorting bands more than about 350 bases from the primer. This effect is caused by a reaction between the glycerol and the boric acid in the gel buffer. Glycerol tolerant gels can therefore be made by replacing the boric acid with taurine (Amersham/USB US 71949).

5. Troubleshooting

As with any complex protocol, problems with sequencing do occasionally arise. The first question to ask is whether the problem is template-specific or systematic, by comparing reaction sets from the same gel. Template-specific problems generally have one of the following causes:

- template DNA contaminated with PEG, RNA, or other DNA (Sections 5.4.1.*i* and 5.4.2.*i*)
- insufficient template DNA (Section 5.3.2)
- no primer binding site in template (Section 5.1.1.*ii*)
- strong secondary structure in template (Sections 5.4.1.*i* and 5.5).

Systematic problems generally have one of the following causes:

- defective reagent (Section 5.1.2)
- gel electrophoresis is suboptimal (Sections 5.1.3 and 5.2).

Of course, template problems can affect a whole gel if there was a systematic problem with the production of a whole batch of templates. A more detailed analysis of possible problems is presented below, categorized according to the symptoms that they present.

5.1 No bands visible

5.1.1 Poor template DNA

There are two main problems with the template DNA that can lead to no sequence ladder being visible.

i. No DNA

Check that there is some DNA in the template preparation by running some on an agarose gel. Remember that single-stranded DNA has a higher mobility than double-stranded DNA of the same length. When preparing single-stranded DNA from phagemids, look for a dense but not confluent culture after overnight growth with the helper phage. A confluent culture may indicate that no phage is being produced, and a very low density culture may indicate that the helper phage cannot infect the bacteria, possibly due to loss of the F′ episome. Many common host strains of *E. coli* have a selection for the F′, e.g. antibiotic resistance, and selection for such a marker can be used to ensure that all stocks for phagemid production carry an F′.

ii. No primer site

If there is no site in the template complementary to the primer, then it cannot anneal. This generally gives no sequence bands at all, sometimes there is weak annealing to another site almost complementary to the primer, giving a very faint background ladder. Apart from the obvious errors in subcloning and synthesizing the primers, this problem can also arise when using nested deletions (see Section 6.3.2.*iii*). If the exonuclease is not properly prevented from digesting the vector, then some bidirectional deletion may occur. If this does not progress very far into the vector sequence then the priming site may be deleted without affecting the sequences essential for propagation of the template, giving DNA which cannot be sequenced. If exonuclease III is used, the deletions in each direction may be of very different lengths, so even a long deletion does not necessarily have the priming site, even if it can be propagated. One particular cause of this problem is contamination of the deletion reaction with single-strand specific nucleases: these will gradually remove the protection of the 3′ overhang, allowing bidirectional deletion. Try phenol extracting the DNA before making the deletions.

5.1.2 Defective reagent

If any one of the key reagents is defective, it is impossible to get good sequence. A likely candidate is the radiolabel. In a busy sequencing laboratory, the label will be thawed and re-frozen quite frequently. This results in chemical degradation of the labelled nucleotide so that it may no longer be suitable for use, even within a single half-life for the ^{35}S. This also applies to the other nucleotide mixes, but their higher concentration makes this less of a problem. Another candidate is the enzyme. The DNA polymerase is not stable for very long at room temperature, and should be diluted in ice-cold enzyme dilution buffer just before use. Apart from these two, there are so many reagents required for sequencing that it is not cost-effective to try to determine which one is at fault. Try sequencing a control template and if this fails then make up all the reagents fresh or buy a good kit (e.g. Sequenase v2.0 Amersham/USB Cat. No. US 70770, see Section 6.1).

5.1.3 Radioactivity remains at top of gel

If most of the radiolabel appears not to enter the gel very far, if at all, the most likely explanation is that the samples have not been properly denatured before loading the gel. The samples are consequently loaded as hybrids of much higher molecular weight which have very low mobility on sequencing gels. Allowing the sample to re-anneal before loading has a similar effect. Just before loading the samples heat them to 85–95 °C for at least two minutes and then cool rapidly on ice.

5.2 Bands not sharp and well-defined

5.2.1 Bands smeared

Band smearing can be caused by using contaminated template DNA or by a gel problem. These problems can severely reduce the amount of sequence information that can be read from a gel. If the bands are smeared in some sets of GATC lanes on a gel and not others, then try again with a fresh preparation of template. Gel problems are harder to solve as they depend on the precise conditions used. In general, sequencing gels should be run at 50–55 °C and dried at only moderate temperature (70–80 °C) on a vacuum drier. Check these temperatures and try using fresh acrylamide solutions.

5.2.2 Bands wavy

While pre-running the sequencing gel, urea diffuses out of the gel into the wells. If this is not carefully removed by washing the wells using a Pasteur pipette, then gel loading and the entry of the sample into the gel is affected. The characteristic result is a wavy appearance to the bands.

5.3 Some sequence faint

If the sequence is very faint over the whole gel, then this probably indicates one of the problems discussed in Section 5.1. This section deals with problems that can result in some regions or lanes giving faint sequence while others are satisfactory.

5.3.1 Sequence faint near the primer

i. Ratio of ddNTP to dNTP is too low

A low ddNTP/dNTP ratio will increase the average length of the terminated chains and so make bands corresponding to the shorter terminated chains relatively faint. If using 'home-made' termination mixes, try increasing the ddNTP/dNTP ratio by increasing the amount of ddNTP. This problem may not affect all four reactions equally, if there has been an error in making up one termination mix, or if the sequence has heavily biased nucleotide usage (see also Section 5.3.3).

ii. Insufficient annealed template

If there is insufficient template or primer in a reaction, then the average chain length after the labelling reaction will be somewhat longer. This means that after the termination reaction there will be very few sequence-specific terminations close to the primer and so bands corresponding to sequence close to the primer will be very faint and are often unreadable. This can be remedied in several different ways. The simplest solution is to use more template DNA and to check the primer concentration to ensure that enough primer is being used. Using less than 0.5 pmol of either template or primer can cause these problems and at least 2 pmol per reaction can normally be used. There are also two possible changes to the reaction conditions which will reduce the average length of the terminated chains, and so enhance the sequence close to the primer, although it is still important to have sufficient template DNA and primer for these methods.

The average length of the synthesized strands can also be reduced by reducing the nucleotide concentration in the labelling reaction. This is done by using dNTP concentrate diluted 10 × or 20 × instead of 5 × in the labelling reaction. Of course reducing the length of the DNA synthesized at this stage reduces the amount of label incorporated in to the reactions and so the resulting bands may be rather fainter, requiring correspondingly longer exposure times.

The addition of Mn^{2+} to the sequencing reactions described above reduces the average length of chain extension in the termination reaction and so enhances sequence close to the primer. Of course this method reduces the intensity of higher molecular weight bands and so sequence more than about 250 bases from the primer cannot normally be read using this method. Simply add $MnCl_2$ to the labelling reaction to a final concentration of 5–10 mM.

5.3.2 Sequence fades away from the primer

Using the protocols described in this chapter, the sequence-specific bands should begin to fade at 600–800 bases from the primer. This depends on the concentrations of template DNA, primer, and label, amongst other factors. If the sequence fades earlier than this, the most likely problems are that:

- insufficient template DNA has been used
- the ddNTP/dNTP ratio is too high
- concentration of dNTP, probably the radiolabel (dATP) is too low.

In any case, the sequence can probably be read with a longer autoradiography exposure. Most gel systems cannot resolve bands beyond the nominal 600–800 base length of these protocols, but it is possible to modify the reaction conditions to give longer synthesized chains if necessary. This is done by using undiluted dNTP concentrate in the labelling reaction instead of 5× diluted concentrate and increasing the concentration of labelled nucleotide fivefold as well. This increase in concentration of all the nucleotides increases the length of DNA synthesized in the labelling reaction and so intensifies bands in the 200–800 base range. Of course sequences close to the primer cannot be read when using these conditions.

5.3.3 Sequence faint in one lane

This may be due to biased nucleotide usage in the template, for example the presence of a poly(A) tract in a cDNA clone. Sequencing through such a sequence will specifically deplete the reaction mix of the complementary nucleotide (dTTP), resulting in faint sequence in this lane beyond the poly(A) region. Try adding extra nucleotide to the labelling mix to compensate, e.g. 1 μl of 0.5 M dTTP. In my experience, poly(A) tracts tend to make sequence faint or ambiguous for up to 20 nucleotides beyond the poly(A).

5.4 Bands in more than one lane
5.4.1 Band across all four lanes
i. Problems with the template
The presence of a band across all four lanes shows that the polymerase has stalled at this point and stopped synthesizing the labelled strand. Termination

is thus independent of the presence of specific ddNTPs and so a band is seen in each lane. This phenomenon is called pausing. The most likely cause is a bad preparation of template DNA, e.g. contaminated with PEG or RNA. This problem is generally much worse with double-stranded DNA templates than with single-stranded ones, and is particularly a problem when using dITP, which prohibits the use of dITP for general sequencing, despite its advantages in eliminating compressions. Pausing is also sequence-dependent. Regions of strong secondary structure may reproducibly show pausing even with clean templates. Sequencing the other strand will usually overcome this problem. If necessary, add 0.5 µg of single-stranded DNA binding protein (SSB protein, Amersham/USB Cat. No. US 70032) to the labelling reaction. This must be inactivated before running the gel by adding 0.1 µg proteinase K and incubating at 65 °C for 20 minutes after adding the stop solution.

ii. Problems with the reactions

Bands can also be seen across all lanes as a consequence of certain errors in performing the reactions, or use of inappropriate reaction conditions. Specifically, the labelling reaction should not be carried out above 20 °C, or pauses may be observed close to the primer. Removing the reaction from ice to room temperature at the start of the labelling reaction, as recommended in *Protocol 4*, should satisfy this condition. Similarly, the termination reactions should not be run colder than 37 °C. Make sure that the termination mix and container for it are pre-warmed for at least one minute before adding the labelling reaction. Termination reactions are best run at 40–45 °C and can be run up to 50 °C if necessary. Glycerol can be added to the reactions to stabilize the T7 DNA polymerase at higher temperatures, but this requires modifications to the gel electrophoresis conditions (see Section 4.3). In general, if pauses are a problem, try again with a fresh batch of RNA-free template, following the advice on temperatures above, and making sure that the reaction components are thoroughly mixed at each stage, without introducing bubbles. Keeping the reaction times for the labelling and termination reactions to five minutes or less may also help.

5.4.2 Bands across two or three lanes

Compressions are regions in which the secondary structure of the DNA results in anomalous migration of the labelled DNA during gel electrophoresis. This can result in two or three bands appearing to have similar mobility, but this is just one possible manifestation of a more general problem which is discussed in Section 5.5 (compressions). Other explanations for finding bands of the same mobility in two or three lanes are described here.

i. Contaminated template

If the template DNA used contains more than one species of DNA to which the primer can anneal, then the observed sequence ladder will be a composite of the sequences of each template. This is also true in the less likely case of contaminated primer. Whether this is a problem depends on the molar ratios of the two templates, but in general the best answer is to prepare fresh template DNA. Note that the contaminated DNA may be the result of a spontaneous deletion during growth of the template. This is a common problem with large insertions into M13 vectors. Try growing the phage for a shorter time, e.g. five hours.

ii. Extra primer site
If the template contains more than one site to which the primer can anneal then the sequence ladder will be a composite of the sequence from each site. This may occasionally occur as an artefact of cloning and subcloning. Alternatively, the primers used may anneal to a repeated sequence in the clone. This problem is exacerbated by using the primer at too high a concentration. This will increase the risk of primers binding to sites to which they are nearly complementary and so giving 'ghost' bands.

5.5 Compressions
5.5.1 Causes
Regions of the gel in which the bands run close together so that their order cannot be determined are known as compressions. In extreme cases bands can run with their relative order reversed, leading to errors in the deduced sequence. Compressions can be very hard to spot and are probably the most common cause of error in published sequences. The most reliable way of identifying these potential errors is to sequence both strands in their entirety and look for discrepancies in the two sequences (see Section 6.4.2). Compressions are caused by strong secondary structure in the radiolabelled strand, usually GC-rich inverted repeats near the compression.

5.5.2 Resolving compressions
This problem can be relieved by substituting dITP for the dGTP in the labelling and termination reactions (*Protocol 10*) as I–C base pairs are less stable than G–C ones.

Protocol 10. Sequencing with ITP

Reagents
- dITP
- 5 × dNTP concentrate for dITP: 15 μM dITP, 7.5 μM dCTP, 7.5 μM dTTP

Method
Follow *Protocol 4* with the following changes.

1. Labelling reaction. Replace the 5 × dNTP concentrate with 5 × dNTP

---

concentrate for dITP. This replaces the 5 × dNTP concentrate and so should be diluted 5 × in water before use in the labelling reaction.

2. (Optional.) Add 0.05 μl sequencing grade pyrophosphatase (USB Cat. No. 70950) per reaction. It is generally convenient to dilute this enzyme together with the DNA polymerase.

3. Termination reactions. In place of the ddA, ddT, and ddC termination mixes for dGTP sequencing use mixes in which the 80 μM dGTP is replaced with 80 μM dITP. For the ddG mix replace the 80 μM dGTP with 160 μM dITP, and the 8 μM ddGTP with 1.6 μM ddGTP.

Unfortunately dITP reactions cannot be read as far as dGTP ones, and are more liable to pausing (Section 5.4.1) and other artefacts, so dITP is not recommended for use in routine sequencing. Other base analogues such as 7-deaza-dGTP are available for the same purpose, but I have never come across a compression that cannot be unambiguously resolved by using dITP sequencing in both directions across the compression.

When using dITP, make sure that the labelling and termination reactions are not allowed to run beyond five minutes each. As well as increasing the risk of pauses (Section 5.4.1) this may also lead to a reduction in intensity of some bands. This is due to pyrophosphorolysis, a slow degradation of the DNA which is a particular problem with dITP reactions. This can also be overcome by adding pyrophosphatase to the sequencing reaction, as described in *Protocol 10*.

Since dITP reactions do not have the same mobility on a sequencing gel, it is not particularly useful to run dGTP and dITP reactions of the same template alongside each other. For most routine sequencing, run dGTP reactions only, then go back and use dITP sequencing to resolve ambiguities (see Section 6.4)

6. Sequencing strategies

6.1 Getting started

All the reagents required for sequencing can be purchased separately, and the various nucleotide mixes and buffers made up from their constituents. For a laboratory performing a lot of sequencing this can result in a considerable cost saving. However, for someone starting a sequencing project, I strongly recommend buying a sequencing kit. This ensures that all the reagents are optimized, removing one set of potential problems, and leaving the researcher to concentrate on the quality of template DNA and actually performing the reactions. There are a number of commercially available sequencing kits, but I can highly recommend the Sequenase v2.0 kit (Amersham/USB Cat. No. US 70770). This kit is based on the Sequenase v2.0 enzyme, which is a modified T7 DNA polymerase which lacks 3′–5′ exonuclease activity. The kit includes all the reagents necessary for dGTP and dITP sequencing together with comprehensive protocols. Many laboratories use this kit for all their sequencing, considering the time saved by using pre-prepared reagents of known reliability to be worth the extra cost. Others replace the kit reagents with 'home-made' preparations as they run out. This allows the quality of each one to be checked as it is introduced.

Before starting sequencing it is important to have a clear strategy that fits the requirements of the project. For this purpose there are two classes of sequencing project: confirmatory sequencing to map and identify mutations, and to check the results of various DNA manipulations, and *de novo* sequencing in which the aim is to sequence a substantial region of DNA for the first time.

6.2 Confirmatory sequencing

Confirmatory sequencing often requires just a single set of reactions to determine the sequence of the few bases of interest, e.g. sequencing the junction between two ligated sequences to check that the reading frame is conserved. This can easily be performed either by subcloning so that the region of interest is close enough to vector sequence to use a 'universal' primer, or else by synthesizing a primer which will anneal to a region about 80 bases away from the region.

Rather than just a few bases, sometimes the sequence of a substantial region of DNA must be checked. An example of this might be sequencing mutant alleles of a gene for which the wild-type sequence is already known. This is most easily done by synthesizing primers which will anneal 300–400 bases apart over the whole region. A single preparation of template DNA then allows all of the sequence information to be determined in a set of reactions using each of the primers in parallel. This approach is particularly efficient if a number of alleles of the same gene need to be sequenced as the same primers can be used for each.

6.3 Sequencing entire clones of DNA

The aim of *de novo* sequencing projects is to determine the entire sequence of a DNA clone, for example a cDNA. These are usually too long for their sequence to be determined by sequencing in from the ends of the clone using primers complementary to the vector. It is therefore necessary to have a strategy which will give sequence information from the middle of a clone. Such strategies fall into two classes: random and directed. Random sequencing generally gives about 90% of the sequence quite quickly and easily, but, depending on the method used, may never give the entire sequence. The sequence is usually completed by another method (see Section 6.4). Directed sequencing is slower at the beginning, owing to the requirement to construct appropriate subclones, but the sequences are easier to compile and give a

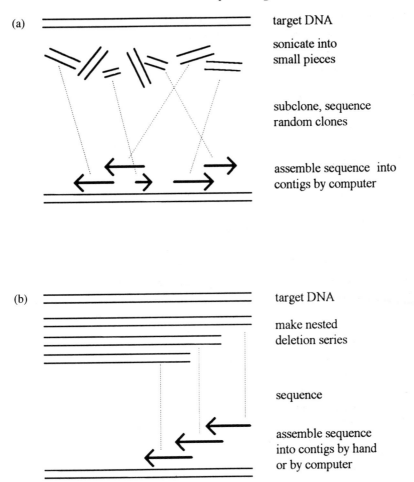

(a) target DNA

sonicate into
small pieces

subclone, sequence
random clones

assemble sequence into
contigs by computer

(b) target DNA

make nested
deletion series

sequence

assemble sequence
into contigs by hand
or by computer

Figure 1. (a) Random sequencing using sonication; (b) directed sequencing with exo-nuclease III.

higher proportion of the total sequence. In either case it is worth first sub-cloning the DNA of interest into pieces of manageable size, i.e. not much more than 5 kb.

6.3.1 Random sequencing

Random or 'shotgun' sequencing involves sequencing subclones of random fragments of the DNA of interest. These random sequence fragments are then compiled into contigs (Section 6.4) and so the entire sequence is determined. There are several methods for generating short subclones from a longer clone, which can then be used as sequencing templates using primers complementary to the vector. In all these methods, purify the DNA fragment of interest away from vector sequences, or you may find yourself sequencing fragments of vector.

i. Restriction endonuclease digestion

Digestion of the DNA of interest with 'four-cutter' restriction enzymes such as *Sau*3A or *Taq*I gives short fragments that are mostly of a suitable size for sequencing. These can easily be subcloned by virtue of their cohesive ends and so sequence information can be obtained very quickly. Subcloning these fragments can give artefacts where two restriction fragments which were not adjacent in the original clone are ligated together. When assembling the sequence information it is therefore important to treat each such fragment separately. They can easily be identified when reading the gel as there will be a site for the enzyme at the junction between each fragment (e.g. GATC for *Sau*3A). It is clearly necessary to use at least two enzymes separately in order to be able to read over the restriction sites in the original clone and so compile the fragments into a contig. The main drawback of this approach is that the fragments generated are not genuinely random. Since small fragments are subcloned more efficiently than larger ones, the small fragments tend to be sequenced many times over, while the large ones may not be cloned at all. There are ways of producing subclones that are genuinely random, but they usually give more problems in subcloning. Two such methods are described below.

ii. Sonication

Sonication can be used to shear DNA into the approximately 500 bp fragments that are ideal for templates. Since this process is very efficient and genuinely random (but see below), this is the method of choice for generating clones for shotgun sequencing. Although the procedure is conceptually simple, a few precautions should be observed to avoid some potential pitfalls.

(a) Purify the DNA of interest away from vector DNA, for example by preparative gel electrophoresis, to avoid sequencing vector DNA.

(b) Ligate the insert DNA to itself to form concatemers and circles before ligating. Without this step the DNA will tend to shear in the middle and together with the presence of a pre-existing break at the ends of the DNA this will unfavourably bias the distribution of subclones.

(c) To eliminate contamination of the DNA from a dirty probe use a cup-horn sonicator, or else make sure that the sonicator probe is scrupulously clean.

(d) Sonicate in several short bursts, cooling between each, to avoid denaturing the DNA by overheating. This problem can also be minimized by using a relatively high ionic strength buffer (e.g. 100–500 mM Na^+).

(e) Aim to shear to fragments of 500–1000 bases with only a small proportion shorter than this. Over-sonicating typically gives fragments of 300–500 bp, but these are difficult to subclone.

(f) Gel purify fragments of 500–1000 bp and subclone. This size selection ensures that sequence data is obtained only from single fragments rather than two shorter ones ligated together.

(g) Sonicated DNA has frayed ends that must be repaired before subcloning, for example using T4 DNA polymerase.

Note that the fragments generated by this procedure will not uniformly represent the starting DNA. Specifically, fragments from the ends of the DNA of interest will be underrepresented. Following self-ligation, the original ends of the DNA are partly or largely represented as inverted repeats, which are very inefficiently cloned by standard methods. This underrepresentation is not normally a problem as the ends can readily be sequenced from the original clone without sonication.

iii. DNase I digestion

In the presence of Mg^{2+}, pancreatic DNase I nicks double-stranded DNA but in the presence of Mn^{2+} it introduces double-stand breaks. This activity can be used to generate random fragments, but is a much less efficient process than sonication and harder to control. A detailed protocol can be found in ref. 3.

iv. Transposon-facilitated DNA sequencing

Integration of bacterial transposons into the target DNA allows the sequence of the DNA around the insertion site to be determined by using primers complementary to the known sequences at the ends of the transposon. Insertion of the transposon is essentially random, but the insertions can easily be mapped, for example by PCR between the ends of the vector and the ends of the transposon. One such method, using the $\gamma\delta$ transposon, is described in ref. 4.

6.3.2 Directed sequencing

There are several diverse strategies by which DNA sequence can be systematically obtained. They all depend on getting sequence from fragments of DNA of known location and orientation relative to the initial clone. This makes compiling the sequence information much easier. The strategies vary in how internal DNA sequence from the original clone is made accessible for sequencing.

i. Subcloning

Working from an accurate restriction map, it is often possible to subclone a set of specific restriction fragments so that much of the clone of interest becomes accessible to sequencing from 'universal' primers. Unfortunately, this is rather labour intensive, as each fragment must be purified and subcloned separately. It is also not normally possible to sequence the whole of a large piece of DNA by this method—there will normally be gaps where there are no convenient restriction sites (see Section 6.4).

ii. Synthesizing primers

Given unlimited access to an oligonucleotide synthesizer, it is possible to 'walk' along the DNA, sequencing in from the end, then using this sequence information to design a new primer, which is then used to obtain more sequence, and so on. This strategy is in fact rather slow, as only one reaction can be performed from each end, and a new oligo must then be synthesized. If oligos are easily obtained, this is one of the least labour intensive methods of sequencing. It can be made much faster by using one of the other strategies, for example subcloning or sonication to generate some more start points for the 'walking' process.

iii. Nested deletions using exonuclease III

The aim of making a deletion series for sequencing is to remove varying amounts of DNA from the end of the clone, bringing new sequence within range of sequencing reactions using a 'universal' primer (see *Figure 1*). Such deletion series can be constructed by a number of methods, of which by far the best uses exonuclease III. Other methods include the use of *Bal*31 or DNase I. The principle advantage of the Exo III method is that no subcloning is required after the exonuclease reaction as the vector DNA can be protected from digestion. Exo III also digests the DNA at a very predictable and reproducible rate, which means that each time point in the reaction is a group of deletions of very similar end-points.

Exonuclease III removes nucleotides from double-stranded DNA at recessed or blunt 3' ends but not from protruding 3' ends. This allows the vector DNA including the primer binding site to be protected from the exonuclease while the DNA of interest is progressively digested. This requires that the target DNA be cut between the priming site and the insert with a restriction enzyme which gives resistant ends, e.g. *Kpn*I, or *Sac*I and one which gives exonuclease sensitive ends, such that the vector DNA is protected and the insert DNA is not, i.e. with the resistant site between the priming site and the sensitive site.

There are a number of commercially available kits for creating nested deletions with exonuclease III. These may be worth trying in case of difficulty with the procedure in *Protocol 11*.

Protocol 11. Creating nested deletions with exonuclease III

Reagents

- Exonuclease III buffer; 66 mM Tris–HCl pH 8.0, 6.6 mM $MgCl_2$
- Exonuclease III (e.g. Pharmacia 27–0874)
- 10 × S1 buffer: 2.3 M NaCl, 300 mM potassium acetate, 2 mM $ZnSO_4$, 50% glycerol
- S1 reaction mix: 0.3 U/μl nuclease S1 in 1 × S1 buffer
- Nuclease S1 (e.g. Pharmacia 27–0920)
- S1 stop solution: 300 mM Tris base, 50 mM EDTA pH 8.0
- Klenow reaction buffer: 200 mM $MgCl_2$, 10 mM Tris–HCl pH 7.5
- Klenow enzyme (e.g. Pharmacia 27–0928)
- 1 mM dNTP mix: 1 mM dATP, 1 mM dCTP, 1 mM dGTP, 1 mM dTTP

A. *Preparing the substrate DNA*

1. Prepare highly purified DNA. As exonuclease III can digest from single-strand nicks in the DNA as well as from the ends, it is important that the DNA substrate is largely nick-free. This is true for DNA purified on a CsCl gradient but not usually true for mini-prep DNA. It is also preferable to remove small fragments of DNA and RNA from the substrate as these may affect the rate of digestion.

2. Digest 5–10 μg of DNA with an enzyme generating Exo III sensitive ends (blunt or 3' recessed) that cuts between the target DNA and the primer site, and with an enzyme giving Exo III resistant ends (3' protruding) that cuts between the sensitive site and the primer site. These restriction enzymes must not cut elsewhere in the sequence. Note that restriction enzymes do not always efficiently cut two adjacent sites.

3. The DNA is purified by phenol/chloroform extraction and then ethanol precipitated.

B. *Exonuclease reaction*

1. Resuspend the DNA in 60 μl of exonuclease III buffer.

2. Pre-warm the DNA solution to 37 °C.

3. Remove 2.5 μl of the solution into S1 reaction mixture as t = 0 control.

4. Add 150 U of exonuclease III to remainder of DNA.

5. At 30 sec intervals, remove 2.5 μl aliquots of Exo III reaction into 7.5 μl S1 reaction mix. These conditions remove approximately 100 nucleotides per time point. This can be adjusted by varying the time interval between the time points.

C. *S1 or mung bean nuclease*

1. Once all the Exo III time points have been taken, incubate the time points, which are now S1 reactions, at 30 °C for 30 min. Add 1 μl of S1 stop solution to each sample to stop the S1 reaction.

2. Incubate at 70 °C for 10 min to heat inactivate the enzymes.

3. Analyse the reaction products by agarose gel electrophoresis of aliquots of some of the reactions. Choose appropriate time points with which to continue. This typically means aiming for deletions with endpoints at 200 base intervals.

D. *Recircularize*

Nuclease S1 does not give a high proportion of blunt ends, so an end-filling reaction using Klenow enzyme or T4 DNA polymerase increases the efficiency with which deleted clones are recovered. Both these enzymes also have a 3'–5' exonuclease activity which will remove any remaining

Protocol 11. *Continued*

protruding 3′ termini from the DNA. For example using the Klenow fragment of *E. coli* DNA polymerase:

1. Add 0.1 U of Klenow enzyme in 1 μl of Klenow reaction mix to each
   time point and incubate at 37 °C for 5 min.

2. Add 1 μl of 1 mM dNTP mix and incubate at room temperature for 15
   min.

3. The DNA can now be recircularized by ligation and used to transform
   an appropriate strain of *E. coli*.

4. Double-stranded DNA can be isolated from the resulting colonies and
   analysed by restriction mapping or sequencing, but it is often more
   efficient to prepare single-stranded DNA (*Protocol 2*) and to sequence
   this directly without further analysis.

iv. Factors affecting the choice of strategy

The choice of strategy depends largely on personal preference and the expertise and facilities available, but also to an extent on the likely structure of the
DNA sequence. The principle factors are discussed below:

(a) If the target DNA is likely to have repetitive sequences within it, then
assembling the accumulated sequence into contigs (see Section 6.4.1) will
be very difficult by hand and computer programs may produce incorrect
contigs. The only way around this is to know where each piece of sequence
information comes from relative to the target DNA. This requires a
directed approach. Furthermore, oligonucleotides designed to anneal to
one part of the target DNA may in fact anneal to another and so can only
be used with great caution. This type of target DNA is best sequenced by
using subcloning or generating nested deletions to produce the templates.

(b) Shotgun sequencing requires the use of a computer to compile the contigs. Although it is possible to compile contigs from directed sequencing
strategies by hand, subsequent analysis of the sequence will need com-
puting facilities and the wide availability of such software means that all
sequencing laboratories should have access to them.

(c) Sequencing by progressively synthesizing oligonucleotides obviously requires ready access to an oligo synthesizer. In fact, one or two oligos are
often synthesized to finish the sequencing (Section 6.4). Fortunately there
are a number of companies which will rapidly synthesize custom oligos on
request.

(d) A number of the strategies generate material that may be useful for later
work. For example, oligos complementary to the sequence may be useful
for PCR amplification of mutant alleles, or nested deletions may be useful

for studying functional domains within the target DNA. Since generating this material is a significant part of the effort and expense involved in a sequencing project, consideration of the likely future direction of the work may influence the choice of strategy.

6.4 Assembling contigs and finishing the sequence

As new sequence information is acquired, it needs to be assembled into longer tracts of contiguous sequence (contigs) based on the overlaps between the sequence information obtained from individual reactions and gels. This can sometimes be performed by hand, depending on the sequencing strategy used, but is always quicker and easier by computer. There are a range of software packages for microcomputers designed to do this and other sequence analysis tasks. You may also have access to a main-frame with such software. Consideration of such software is beyond the scope of this chapter.

6.4.1 Joining the contigs

The sequencing strategies described will eventually give the entire sequence of the target DNA, but it is often more efficient to consider additional approaches when the sequence is nearly completed. For example, certain regions of the target DNA may be hard to clone by shotgun methods, unrepresented in a deletion series, or may contain a repeat, etc. It is therefore often more efficient to finish the sequence by synthesizing oligos to extend each contig in the hope that these new sequences will overlap with another contig and so allow the whole sequence to be compiled into a single contig. Comparing the computer predicted restriction maps of the contigs with the restriction map of the target DNA may indicate the orientation and position of the contigs with respect to the target DNA. It may also be possible to subclone specific fragments so as to sequence DNA that has not been sequenced by the original strategy.

6.4.2 Maximizing the accuracy of the finished sequence

i. Sequencing both strands

Completed sequence should have an error rate of less than 0.1%. For the final sequence to be of this high quality it is essential to determine the whole sequence in both directions. This is the only reliable way to avoid errors due to sequence secondary structure (see Section 5.5). A number of published sequences contain significant errors through not sticking rigidly to this rule! Just as the contigs can be joined by sequencing from oligonucleotides designed for the purpose, this can also be used to sequence DNA whose sequence has been determined in one direction only.

ii. Proof-reading and resolving discrepancies

Most of the target DNA will normally have been sequenced several times over from independent templates, although this depends on the strategy used.

In the process of aligning this sequence information into contigs, discrepancies between the sequences will become obvious. These need to be checked carefully against the original gels. Common errors include miscounting the number of bases in a homopolymeric tract, especially near the top of a gel. Where this proof-reading does not resolve the discrepancy, re-sequence the templates concerned with dITP to eliminate compressions (Section 5.5 and *Protocol 10*).

iii. cDNAs

Sometimes the target DNA will have a predictable sequence structure which can be used to check the sequence of the contig. For example, cDNA clones will normally have a long open reading frame. The presence of two overlapping open reading frames may suggest a frame-shift in the region of overlap. This is almost always due to the insertion or omission of a base while reading in the sequence, possibly due to a compression. Very occasionally the explanation is that the cDNA clone being used itself has a frame-shift mutation in it, probably as an artefact of cDNA cloning. Of course a very small proportion of sequences may genuinely have frame-shifts as a means of regulating protein expression levels, or 'stop' codons which in fact encode selenocysteine etc. The 5' end of a cDNA clone cannot be said to correspond to the 5' end of the message without further analysis, such as primer extension (see Chapter 5). In fact, depending on the library used, the 5' end of a cDNA may well have some sort of cloning artefact, for example a rearrangement or the addition of some non-coded bases as linkers or tailing. If the sequence in this region is of interest then it is well worth cloning and sequencing the corresponding genomic DNA. Sequencing the genomic region will also allow the identification of introns and promoter sequences and will serve as a check on the cDNA sequence for both proof-reading and cloning errors.

References

1. Maxam, A. M. and Gilbert, W. (1980). In *Methods in enzymology*, Vol. 65, pp. 499–500 Academic Press, London.
2. Sanger, F., Nicklen, S., and Coulson, A. R. (1977). *Proc. Natl. Acad. Sci. USA*, **74**, 5463.
3. Sambrook, J., Fritsch, E. F., and Maniatis, T. (ed.) (1989). *Molecular cloning, a laboratory manual* (2nd edn). Cold Spring Harbor Press, Cold Spring Harbor, NY.
4. Strathmann, M., Hamilton, B. A., Mayeda, C. A., Simon, M. I., Meyerowitz, E. M., and Palazzolo, M. J. (1991). *Proc. Natl. Acad. Sci. USA*, **88**, 1247.

A1

List of suppliers

Amersham International plc., White Lion Road, Amersham, Bucks HP7 9LL, UK; 2636 S. Clearbrook Drive, Arlington Heights, Il 60005, USA.

Amicon, Upper Mill, Stonehouse, Gloucester, GL10 2BJ, UK; 17 Cherry Hill Drive, Danvers, MA 01923, USA.

Anderman & Company Ltd., 145 London Road, Kingston Upon Thames, Surrey, KT2 6NH, UK.

Applied Biosystems Inc., 850 Lincoln Centre Drive, Foster City, CA 94404, USA. Biotech Instruments Ltd., Unit A, Caxton Hill Extension Road, Caxton Hill, Hertford, SG13 7LS, UK.

Beckman Instruments Inc., 2500 Harbour Boulevard, P.O. Box 3100, Fullerton, CA 92634, USA; Progress Road, Sands Industrial Estate, High Wycombe, HP12 4JL, UK.

Bio-Rad Laboratories Ltd., Bio-Rad House, Maylands Avenue, Hemel Hempstead, Herts, HP2 7TD, UK; 1414 Harbor Way South, Richmond, CA 94804, USA.

Boehringer-Mannheim GmbH., Postfact 310120, D-6800, Mannheim 31, West Germany; P.O. Box 50816, Indianapolis, IN 46250, USA; Bell Lane, Lewes, BN17 1LG, UK.

Calbiochem-Novabiochem, Boulevard Industrial Park, Padge Road, Beeston, Nottingham, NG9 2JR, UK.

Corning, UK distributor—Bibby Sterilin Ltd., Tilling Drive, Stone, Staffs., ST15 0SA, UK.

Difco Laboratories, P.O. Box 14B, Central Avenue, East Molesey, Surrey, KT8 0SE, UK.

Dynal Ltd., Station House, 26 Grove Street, New Ferry, Wirral, L62 5AZ, UK.

Eastman Kodak, Acorn Field Road, Knowsley Industrial Park North, Liverpool, L33 72X, UK; Rochester, New York 14650, USA.

Falcon, UK distributor—Becton Dickinson UK Ltd., Between Towns Road, Cowley, Oxford, OX4 3LY, UK.

Genetic Research Instruments, Gene House, Dunmow Road, Felstead, Dunmow, Essex, CM6 3LD, UK.

Gibco BRL Life Technologies Inc., Grand Island, NT, USA; P.O. Box 35, Trident House, Renfrew Road, Paisley, PA3 4EF, UK.

Greiner Labortechnik Ltd., Station Road, Cam, Dursley, Glos., GL11 5NS, UK.

Hybaid, 111–113 Waldegrave Road, Teddington, Middlesex, TW11 8LL, UK.

IBI Ltd., 36 Clifton Road, Cambridge, CB1 4ZR, UK; P.O. Box 9558, New Haven, CT 06535, USA.

Merck Ltd., Hunter Boulevard, Magna Park, Lutterworth, Leics., LE17 4XN, UK.

Millipore (UK), The Boulevard, Blackmoor Lane, Watford, Herts., WD1 8YW, UK.

National Diagnostics, Unit 3, Chamberlain Road, Aylesbury, Bucks., HP19 3DY, UK; 1013–1017 Kennedy Boulevard, Manville, NJ 08835–2031, USA

New England Biolabs, 67 Knowl Piece, Wilbury Way, Hitchin, Herts., SG4 0TY, UK; 32 Tozer Road, Beverly, MA 01915–9990, USA.

Oxoid Ltd., Wade Road, Basingstoke, Hants., RG24 0PW, UK.

Perkin Elmer Cetus, Maxwell Road, Beaconsfield, Bucks, HP9 1QA, UK; Main Avenue, Norwalk, CT 06856, USA.

Pharmacia Biotech Ltd., 23 Grosvenor Road, St. Albans, Herts., AL1 3AW, UK; 800 Centennial Avenue, Piscataway, NJ 08854, USA.

Promega Corporation, Epsilon House, Enterprise Road, Chilworth Research Centre, Southampton, SO1 7NS, UK; 2800 Woods Hollow Road, Madison, WI 53711–5399, USA

Qiagen Ltd., Unit 1, Tillingbourne Court, Dorking Business Park, Station Road, Dorking, Surrey, RH4 1HJ.

Sigma Chemical Company Ltd., Fancy Road, Poole, Dorset BH17 7NH, UK; P.O. Box 14508, St. Louis, MO 63178, USA.

Stratagene Ltd., Cambridge Innovation Centre, Cambridge Sciences Park, Milton Road, Cambridge, CB4 4GF, UK; 11099 North Torrey Pines Road, La Jolla, CA 92037, USA.

Index

Index

265

DNA Cloning 3: Complex Genomes

1. Cosmid clones and their application to genome studies
   Alasdair C. Ivens and Peter F. R. Little
2. Chromosome-specific gridded cosmid libraries: construction, handling and use in parallel and integrated mapping
   Dean Nizetic and Hans Lehrach
3. Library construction in P1 phage vectors
   Nat Sternberg
4. Cloning into yeast artificial chromosomes
   Rakesh Anand
5. Amplification of DNA microdissected from mitotic and polytene chromosomes
   Robert D. C. Saunders
6. Databases, computer networks and molecular biology
   Rainer Fuchs and Graham N. Cameron
7. Long range restriction mapping
   Wendy Bickmore
8. Genetic mapping with microsatellites
   Isam S. Naom, Christopher G. Mathew, and Margaret-Mary Town

DNA Cloning 4: Mammalian Systems

1. High efficiency gene transfer into mammalian cells
   Vanessa Chisholm
2. Construction and characterization of vaccinia virus recombinants
   Mike Mackett
3. Vectors based on gene amplification for the expression of cloned genes in mammalian cells
   Christopher R. Bebbington
4. Retroviral vectors
   Anthony M. C. Brown and Joseph P. Dougherty
5. Introduction of cloned DNA into embryonic stem cells
   Amy R. Mohn and Beverly H. Koller
6. Production of transgenic rodents by the microinjection of cloned DNA into fertilized one-cell eggs
   Sarah Jane Waller, Mei-Yin Ho, and David Murphy
7. Genomic and expression analysis of transgenic animals
   Sarah Jane Waller, Judith McNiff Funkhouser, Kum-Fai Chooi, and David Murphy
8. Expression of cloned DNA using a defective Herpes simplex virus (HSV-1) vector system
   Filip Lim, Philip Starr, Song Song, Dean Hartley, Phung Lang, and Alfred I. Geller
9. Adenovirus vectors
   Robert D. Gerard and Robert S. Meidell